The Rise and Fall of the Healthy Factory

The Rise and Fall of the Healthy Factory

The Politics of Industrial Health in Britain, 1914–60

Vicky Long
Lecturer in British History,
Northumbria University, UK

© Vicky Long 2011

All rights reserved. No reproduction, copy or transmission of this publication may be made without written permission.

No portion of this publication may be reproduced, copied or transmitted save with written permission or in accordance with the provisions of the Copyright, Designs and Patents Act 1988, or under the terms of any licence permitting limited copying issued by the Copyright Licensing Agency, Saffron House, 6-10 Kirby Street, London EC1N 8TS.

Any person who does any unauthorized act in relation to this publication may be liable to criminal prosecution and civil claims for damages.

The author has asserted her right to be identified as the author of this work in accordance with the Copyright, Designs and Patents Act 1988.

First published 2011 by
PALGRAVE MACMILLAN

Palgrave Macmillan in the UK is an imprint of Macmillan Publishers Limited, registered in England, company number 785998, of Houndmills, Basingstoke, Hampshire RG21 6XS.

Palgrave Macmillan in the US is a division of St Martin's Press LLC, 175 Fifth Avenue, New York, NY 10010.

Palgrave Macmillan is the global academic imprint of the above companies and has companies and representatives throughout the world.

Palgrave® and Macmillan® are registered trademarks in the United States, the United Kingdom, Europe and other countries.

ISBN: 978–0–230–28371–8 hardback

This book is printed on paper suitable for recycling and made from fully managed and sustained forest sources. Logging, pulping and manufacturing processes are expected to conform to the environmental regulations of the country of origin.

A catalogue record for this book is available from the British Library.

Library of Congress Cataloging-in-Publication Data

Long, Vicky, 1978– , author.
    The Rise and Fall of the Healthy Factory : The Politics of Industrial Health in Britain, 1914–60 / Vicky Long.
        p. cm.
    Includes bibliographical references and index.
    ISBN 978–0–230–28371–8 (hardback)
        1. Industrial hygiene – Great Britain – History – 20th century.
    2. National health insurance – Great Britain – History – 20th century.
    3. World War, 1914–1918 – Great Britain. 4. World War, 1939–1945 – Great Britain. I. Title.

HD7695.L65 2011
363.110941'09041—dc22   2010042396

10 9 8 7 6 5 4 3 2 1
20 19 18 17 16 15 14 13 12 11

Printed and bound in Great Britain by
CPI Antony Rowe, Chippenham and Eastbourne

In memory of my mother, Pat Long

Contents

Abbreviations

BEC	British Employers Confederation
BMA	British Medical Association
HMWC	Health of Munition Workers Committee
IHAC	Industrial Health Advisory Committee
IHRB	Industrial Health Research Board
IHSC	Industrial Health Services Committee
LWMC	Llanelly and District Workmen's Medical Committee
MRC	Modern Records Centre
NHS	National Health Service
PRO	Public Records Office
SMA	Socialist Medical Association
TUC	Trades Union Congress

Acknowledgements

This book draws on research I undertook within the Centre for the History of Medicine at the University of Warwick as research assistant on the Wellcome Trust-funded project, 'The Politics and Practices of Health in Work, 1915–1960' (grant number 076053/Z/04/Z). I am grateful to the Wellcome Trust for its support and would like to express my sincere thanks to the project managers, Hilary Marland and Mathew Thomson. Also my erstwhile PhD supervisors, Hilary and Mathew acted as a sounding board for ideas throughout the project, were enthusiastic and encouraging when I suggested new approaches and provided insightful comments on the manuscript. I am grateful for all their help, advice and intellectual generosity over a number of years. I would also like to thank Jane Adams, Sue Aspinall, Dan O'Connor, Gauri Raje, Molly Rogers, Sheryl Root, Julia Smith, Jonathan Toms and Brooke Whitelaw, who between them made my time at Warwick far more enjoyable than it otherwise would have been.

I would like to express my thanks to the reader for their constructive comments on the initial draft of the manuscript. I completed final revisions to the book while working at the Centre for the History of Science, Technology and Medicine at the University of Manchester and wish to acknowledge the support of the Centre's Director, Michael Worboys. Here, I was fortunate to join an animated and friendly group of post-doctoral researchers who were always happy to discuss ideas, both in the office and down the pub: so thanks to Julie Andrews, Jo Baines, Mike Brown, Kat Foxhall, Val Harrington, Emma Jones, Rob Kirk, Neil Pemberton, Stephanie Snow, Elizabeth Toon, Leucha Veneer and Duncan Wilson. I'd also like to thank Emily Hankin for sharing my fascination with industrial and domestic technologies, and Clare Hickman, who was always happy to discuss histories of health and space on my trips back to the West County.

I carried out much of my research in the Modern Records Centre at the University of Warwick and am very grateful to Christine Woodland, the former archivist at the Centre, for her helpful and informative suggestions at the inception of the project. I would also like to thank Christine's successor, Helen Ford, and other Centre staff, for their assistance throughout and for hosting two workshops linked to the project. Charlotte McCarthy and Sarah Foden facilitated my research at the Boots Archive and Cadbury Archive respectively, and I would like to thank Cadbury UK for permission to reproduce material from its archive. The 'Shredded Wheat' name and image is reproduced with the kind permission of Société des Produits Nestlé S.A. I am also grateful for the assistance rendered by staff at the National Archives and Birmingham Archives and Heritage, and wish to thank Sally and Terry Gannon, who kindly offered me a home from home while I worked at the National Archives.

Many of these chapters originated from conference papers, given to the North American Conference on British Studies in Cincinnati; annual conferences of the Society for the Social History of Medicine in Warwick and Glasgow; the European Social Science History Conference in Amsterdam; the Social History Society Conference in Rotterdam; 'The History of Work, Environment and Health' in Birmingham and two workshops at Warwick: 'Health, Work and Masculinity c. 1800–1950' and 'Beyond the Politics of Motherhood? Women's Health and Bodies 1890–1930'. I am grateful for the constructive comments offered by participants at all these events. The opportunity to present seminar papers at the Centre for the History of Medicine at the University of Warwick, the Centre for the History of Medicine in Ireland at University College Dublin and the Centre for the History of Science, Technology and Medicine at the University of Manchester proved particularly valuable. British Academy Overseas Conference Grants in 2007 and 2010 were gratefully received, enabling me to present aspects of this project at the American Association for the History of Medicine in Montreal and the 4th International Conference on the History of Occupational and Environmental Health in San Francisco.

Finally, I'd like to thank family and friends for their support throughout the project, particularly my partner Sam, who has had to listen to a lot of factory talk over the past few years.

Introduction

Is a healthy workplace good business?

Recent government policy in the field of occupational health has been shaped by Dame Carole Black's *Review of the Health of Britain's Working-Age Population.* Presented jointly to the Departments of Health and Work and Pensions in 2008, Black's Review asserted that 'good health is good business', urging workers and managers to 'recognise not only the importance of preventing ill-health, but also the key role the workplace can play in promoting health and well-being'.[1] Black painted a vision of an ideal workplace which supplied healthy subsidised food within its canteen, offered health screenings and consultations with general practitioners within the workplace and provided incentives to its workforce to participate in exercise by providing onsite sports facilities and by encouraging employees to actively travel to and from work. She urged companies to think carefully when designing jobs, warning that monotonous and repetitive work, a lack of autonomy or discretion over work tasks and a failure to reward effort could worsen employees' health. Seeking to align health concerns and economic goals, Black depicted the implementation of health and well-being initiatives as a sound investment which enabled employers to attract high-calibre staff and reduce the costs of sickness absence.[2]

If workplaces could be reconfigured into sites of health improvement, employment and recuperation need not be inimical. Following this train of logic, Black suggested introducing a 'fit note' that detailed people's working capabilities alongside their impairments. The Department of Work and Pensions implemented this recommendation in April 2010, replacing the medical statement or 'sick note' with the new 'Statement of Fitness for Work'.[3] Black envisaged that 'fit notes' would

1

be complemented by a new 'Fit for Work' service, which would be extended to those on incapacity or other out-of-work benefits. Adopting a multidisciplinary approach, this projected service would provide support for social anxieties such as financial and housing issues alongside national health services such as physiotherapy and talking therapies: Fit for Work Services are currently being piloted in 11 sites across England, Wales and Scotland.[4] Black expressed the hope that her recommendations would deliver higher tax receipts, reduced benefit payment, better workplace productivity and over time 'reduced costs to the NHS'.[5]

The Rise and Fall of the Healthy Factory seeks to illuminate how the historical contexts and processes through which health care services for workers were negotiated and contested over the course of the twentieth century continues to inform government policy in this field, creating precedents which became difficult, if not impossible, to later undo. Viewed through this historical prism, Black's Review reflects how occupational health became irrevocably entangled with political and commercial concerns and was consequently excluded from the National Health Service, leading to the marginalisation of work-related health problems. This book takes as its object of study not the problems of industrial disease, but a broadly conceived model of health which embraced physical and mental well-being in all spheres of life. It traces the origins of the holistic vision of occupational health advocated by Black within her Review back to the years of the First World War, when unprecedented pressures on industrial production forced the government, employers and doctors to rethink the interaction of workplace and worker. Model factories constructed in the aftermath of the War tapped into a vein of utopian thought which reconfigured factory space into a pivotal site in which the relationship between work, leisure, production, consumption, health and citizenship could be reframed.

In 1926, one of the newest buildings in Welwyn Garden City was described by an effusive visitor as 'a beautiful structure, looking in the distance as if it had been carved out of white ivory, and with innumerable windows sparkling like diamonds in the sunshine'. The account described the flowerbeds and lawns surrounding the building, the pleasant and airy rooms that benefited from a design which introduced sunshine and light, the dining room and the ice-cream plant operating in the kitchen.

The edifice in question was the new Shredded Wheat factory. Industrial toil, which already seemed rather incompatible with this setting, was rendered still more incongruous by the description of those employed within the factory as the 'Shredded Wheat family', and the 'community spirit' which pervaded the atmosphere, running counter to concepts of fixed hierarchical relations between employers and employees. Situated within Welwyn Garden City, the factory was seen to exemplify the trend of discarding older premises in urban areas in favour of new custom-built structures in the suburbs. The author of the article asserted that 'a factory may, and should be, as beautiful or at least as pleasing, as a fine school or institution'.[6]

Another writer, Dorothea Proud, envisaged a model factory which would transcend its primary purpose of manufacture to encompass health and welfare initiatives for its employees:

> The ideal factory she opined is neat, even if not beautiful and there are adjacent grounds devoted to gardens and playing-fields. The ventilation, heating, drainage and lighting have all been made the subjects of expert consideration. The floors and walls are constructed, and the machinery placed, so that it is possible to maintain a healthy state of cleanliness...Each workroom has its own first-aid cupboard...Lavatories and sanitary conveniences are private, and yet easily accessible. They are adequate in number and suitable in arrangement. Perhaps there are bathrooms in addition; possibly a swimming-bath and a gymnasium. At any rate, there are cloak-rooms, dining-rooms and rest-rooms such as will please the workers and satisfy their needs.[7]

Writing in 1918, Proud reflected the radical shift which had taken place during the First World War with respect to the form and function of factories. The decision of the government to establish the Health of Munition Workers Committee in 1915 prompted a far-reaching investigation into how lighting, heating, ventilation, nutrition, rest breaks, working hours and recreational facilities affected workers' productivity and health and encouraged state intervention. New professional groups emerged to run health and welfare facilities within factories and established professional groups increasingly engaged with the subject. Industrial health developed as a branch of medicine and in 1920 the first Chair of Industrial Medicine in Britain was established at St Mary's Hospital; by this time there were already six professorships in industrial medicine in America.[8] University departments of industrial health were established at Birmingham in 1935 and at Manchester and Durham in 1945.[9] 'Among all the changes which the Great War has brought in its train,' wrote David Lloyd George, 'none is more significant, and none more likely to have a lasting effect than the revolution in the structure of British industry.'[10]

Thirty-three years later, a government committee was established to report on the possible overlap of service provision between industrial health services and the National Health Service (hereafter NHS) to see if medical manpower for industrial health services should be cut back.[11] Industrial health had moved from the mainstream to beyond the margins, where its very existence was perceived as a threat to the NHS. The findings of the 1972 Safety and Health at Work Committee confirmed its exclusion.[12]

The emergence of the factory

The debate about the place of health facilities within the factory arose partly because the factory system of production was relatively recent and no consensus existed as to what the functions of a factory were. Factories were a product of the Industrial Revolution, a consequence of new technological innovations which emerged first in the textiles industry in the 1770s that made the manufacture of articles by home workers unproductive.[13] The term 'factory' or 'manufactory' referred to buildings in which goods were manufactured but increasingly came to refer to buildings in which production was based around machinery. This explanation is rather misleading in its simplicity, however. Other definitions of the factory move us beyond a basic economic classification, emphasising changing working patterns and hierarchies of power within the workplace. T. G. Geraghty identified four features which distinguish the factory system from previous forms of production: centralised and large-scale production, use of technology to mechanise production processes, discipline and organisation of the workforce and subdivision of labour.[14] In his 1835 account of the British factory system, Andrew Ure criticised the use of the term 'factory' to simply designate an establishment where a number of people worked together towards a common goal. He defined the factory system as 'the combined operation of many orders of work-people, adult and young, in tending with assiduous skill a series of productive machines continuously impelled by a central power'.[15]

Ure was an enthusiastic advocate of what he described as the humane system of factory production, but his fervour was not shared by all. As the factory system of production spread throughout Britain in the late eighteenth century, social change followed in its wake. Growing mechanisation rendered many skilled workers redundant; women and children displaced skilled male workers in some trades such as the cotton industry. Industrialisation tied Britain into global booms and slumps, causing long-term depressions and unemployment. Keen to maximise output and profit, employers subjected factory workers to fixed working hours, discipline and regulations. Through the course of the nineteenth century, skilled workers were losing control over the productive process so that by the end of the century, the erosion of skill and status had reached 'epidemic proportions'.[16]

Factories provided workers with a new environment in which to interact, shaping class identity and forging new forms of industrial relations. Negotiations between employers and employed became increasingly formalised as the trade union movement grew in strength, enhancing its capacity to protect the interests of workers. Organisation among workers in the cotton industry followed rapidly on the heels of the development of the first factories; local associations sought to negotiate wages in Lancashire and Lanarkshire from the late eighteenth century onwards.[17] Through the course of the nineteenth century, trade union organisation became more centralised and unions began to represent workers on a national basis, such as the General Union of Cotton Spinners, which was established in 1829 and which forged links between workers in England, Ireland and Scotland. In 1868, the Trades Union Congress (hereafter TUC) was established to debate issues of interest, and from 1872 to influence government policy through its Parliamentary Committee.[18] Towards the end of the nineteenth century, the trade union movement, which had hitherto been dominated by craft unions, began to attract unskilled workers who were keen to secure immediate improvements in wages and working hours. Dubbed new unionism, this movement was prepared to jettison peaceful settlements to secure its goals. New large unions were formed, such as the United Textile and Factory Workers' Association in 1887 and the Federation of Engineering and Shipbuilding Trades in 1889. The willingness of new unionists to adopt militant strategies led to a resurgence of hostilities between employers and workers.

Efforts to rethink the form and function of the factory emerged almost as soon as the factory system of production began to take hold. Robert Owen had started his working career managing cotton mills of Manchester. In 1800 he purchased the textile factories at New Lanark and strove to create a new model of industry, restricting working hours and inaugurating an extensive welfare scheme for his employees that included housing, education, medical provisions and social and recreational facilities. 'Essentially New Lanark was a repudiation of the city; it was industry in a rural setting' asserted J. F. C. Harrison in his study of Robert Owen.[19] Towards the end of the nineteenth century, a handful of employers began experimenting with ways to humanise their factories. A few companies, driven by a paternalistic vision often informed by Quaker ideals, inaugurated welfare facilities which sought to meet the educative, moral and recreational needs of their employees. The Boots Pure Drug Company, for example, organised works outings for employees from 1888, employed its first welfare supervisor in 1893 and established an athletics club in 1894.[20] These kinds of facilities were the exception rather than the rule, while some welfare provisions, especially profit sharing and benefit schemes, were established by employers hoping to lure workers from trade unions, or else employers strategically targeted young apprenticed workers, women and supervisory staff.[21] In response to campaigns by the Amalgamated Society of Engineers to reduce working hours, Manchester engineering firm Mather and Platt trialled a shorter 48-hour week in the 1890s. The experiment transcended the philanthropic concerns which had hitherto characterised employer intervention in workers' welfare, standing as one of the first investigations into the relationship between productivity and working hours: production levels remained the same while timekeeping improved and absenteeism declined. Previously it had been assumed that the relation between working hours and output was linear and as there was no shortage of labour in the nineteenth century there was little incentive for employers to shorten working hours to benefit their employees.[22] Some firms had begun to introduce supervisors or welfare secretaries and the first conference for such workers was held by Bournville in 1906.

Developments within the state sector lagged behind private initiatives. Nevertheless, towards the end of the nineteenth century the Factory Inspectorate turned its attention to a wider range of issues.[23] Women factory inspectors were first appointed in 1893 after women helped compile evidence for the 1891 to 1893 Royal Commission on Labour. Women inspectors demonstrated more concern with welfare than safety, arguing that women's health was less affected by accidents than general working conditions. Thus women factory inspectors began to turn their attention to the working environment, focusing on lighting, hygiene, ventilation, temperature and the provision of cloakrooms. 'Always in wartime welfare in factories was spoken of as if it were something quite new,' recalled Factory Inspector Hilda Martindale. 'As a matter of fact, the women inspectors, supported by the Home Office, had from the first spread knowledge regarding this movement and inciting employers to follow the example of the forerunners in it.'[24] Alongside the introduction of women, the Factory Inspectorate began to adopt a more medicalised approach. In 1896, the Department was put under the control of Dr Arthur Whitelegge while in 1898 Thomas Legge was appointed as the first Medical Factory Inspector.

By the early twentieth century, the factory had come to assume an iconic status within British culture, symbolising urbanisation, industrial progress, technological innovation and capitalism.[25] As this account will also demonstrate, the modern factory also became emblematic of the new public interest in health improvement demonstrated in the interwar years. While the large factory, with its countervailing horrors and delights, came to dominate the popular imagination, it masked the reality of industrial production: a diverse array of workplaces clustered beneath the heading of factory as defined by legislature, many employing fewer than 26 people.[26] The real problems and needs of these factories were largely neglected until the Second World War.

The history of occupational illness

Legislation governing factory conditions emerged largely in response to developments in a rapidly evolving field and has therefore not surprisingly often been seen as a somewhat piecemeal process, responding to developments in a fragmentary way to protect against the worst hazards or to exclude women and children from particular processes and industries.[27] In his account of public health in Victorian Britain, Anthony Wohl asserted that industrial diseases were viewed as an accepted and inevitable aspect of working life.[28] In a similar vein, Steve Sturdy has argued that as the factory emerged as a key site of production, the contract between employers and employees came to determine the relationship of industrial work. Workers' bodies and the labour they performed were viewed as a form of capital; the market would determine the value of that labour and workers engaged in hazardous or demanding occupations would be paid higher wages to compensate.[29] Legislation in the early nineteenth century was often based on this assumption and focused on regulating and restricting conditions of work for children and women. The Factory Act of 1819 had barred children under the age of nine from working in a mill or cotton factory and restricted children under the age of 16 to 12 hours of work a day, but was widely evaded.

'The factory system is one of absolute tyranny, deeply rooting itself in this land of liberty', wrote Richard Oastler, whose campaign against the evils of child labour had earned him the moniker the Factory King. 'It is a system of impiety and immorality, insinuating its deadly poison.'[30] To Oastler and most of his contemporaries, early twentieth-century visions of beautiful, hygienic factories replete with health and welfare facilities to improve the lives of those that worked within them would have been utterly alien to the vision of the factory which had developed as the factory system of employment and production advanced.

By the 1830s, public disquiet regarding factory work, especially when undertaken by women and children, was rife.[31] Charles Turner Thackrah, a surgeon from Leeds, published the first major account of the medical effects of different branches of industrial work on health in 1831.[32] Thackrah provided his readers with a comparative analysis of the health risks of different occupations, seeking to highlight the pitfalls of modern industrial life and to provide possible solutions to the health risks. Without health, Thackrah argued, the wealth produced through industry was useless to a person.

> We may say that at least 450 persons die annually in the borough of Leeds, from the injurious effects of manufactures, the crowded state of the population, and the consequent bad habits of life...every day we see sacrificed to the artificial state of society one and sometimes two victims...The destruction of 450 persons year by year on the borough of Leeds cannot be considered by any benevolent mind as an insignificant affair. Still less can the impaired health, the lingering ailments, the premature decay, mental and corporeal, of nine-tenths of the survivors, be a subject of indifference.[33]

Government enquiries and legislation followed: the 1833 Factory Act further restricted child labour in all factories, required the fencing of dangerous machinery and established a Factory Inspectorate to enforce legal requirements. The 1891 Factory Act specified particular conditions and welfare provisions for workers within the dangerous trades.

The occupational illnesses and accidents resulting from work which were first identified by Thackrah have subsequently dominated the imagination of historians, leading to a wealth of publications which detail the experience of workers affected by industrial illnesses;[34] the emergence of compensation for workers incapacitated through work;[35] the persistent ill-health suffered by workmen during the decline in mortality rates,[36] and the emergence of the industrial hygiene profession in America.[37] Efforts to promote health by inaugurating medical and welfare facilities linked to or based within the factory, and indeed efforts to transform the environment of the factory to enhance mental and physical health has been dismissed as a strategy adopted by employers to gain control over workforces without providing any real benefits, distracting attention away from the very real hazards of industrial illness.[38] While there is undoubtedly much truth in this observation, it does not explain the emergence and decline of ideas of the healthy factory, nor can it account for the growing interest of the State and trade unions in healthy workplaces.

'Health at work is not a subject that is easily delimited' noted the authors of a 1972 report, as they explained why they had failed to keep within their remit.[39] Not only would I concur with this observation, I would also argue that the inability to confine either the practice or historical analysis of industrial health into watertight compartments is illuminating rather than problematic. To exclude analysis of wages and hours of work on the grounds that it was not a medical issue would be to ignore both the extent to which diet was affected by wages and to lose sight of how trade unions presented traditional industrial relations issues in new garb as health issues. To fail to analyse health schemes beyond the factory gates which were nevertheless targeted at workers would mean a failure to appreciate how ideas about worker-run schemes were contested, with important implications for the trade union movement, and to lose sight of how the development of schemes within the community affected the viability of workplace provisions. Finally, to focus exclusively on health and to exclude from study any material relating to ill-health would be to distort assessment of achievements in the field of health. Consequently, this study provides an analysis of the politics of health promotion within factories but acknowledges that debates about health were intrinsically linked to discussions of illness. It also recognises that many arguments advanced by historians of industrial illnesses and accidents are equally applicable within the field of industrial health. Chris Sellers' argument, for example, that 'questions about knowledge ineluctably engaged questions about responsibility', an insight arrived at in the process of analysing the emergent industrial hygiene profession, remains as compelling in the field of health promotion as it does in his analysis of industrial illness.[40]

The politics of industrial health in the twentieth century

It is a fundamental premise of this account that political and economic developments shaped the history of industrial health provisions and can explain both why industrial health had emerged as a concern of the State in the late nineteenth century and why it took its subsequent course. Developments in the late nineteenth century reflected the growing consensus among the State and the medical profession that levels of industrial production and the health of the labour force were of concern, not only for workers and employers but for the nation.[41] Increasingly, the State perceived the bourgeoning trade union movement, fuelled by the ideas of new unionism, as a powerful, disruptive and potentially threatening force which should be offered some concessions. In the first half of the twentieth century, as John Pickstone has argued, the State intervened to promote the health of the armed forces and workforce, promoting a model of 'productionist medicine'.[42] As nation states sought to maintain their empires and economies, maternal health, infant welfare and the health of the working man rose up the agenda. A great deal of attention has been paid to the emergence of maternal and infant welfare services.[43] However, the health of the working man has been a largely neglected topic, even though industrialisation and the accompanying shift to a factory-based system of production prompted the establishment of nationalised health services. In 1911, the National Health Insurance Act ushered in a set of provisions explicitly designed to preserve the health of workers and safeguard their working capacity, reflecting the new belief that it was beneficial to maintain the productivity of workers, although the medical provisions established under the terms of the Act were based within the community.

If industrialisation had thus driven the State to inaugurate health provisions, it would be logical to expect that any expansion of services would encompass the working environment. International comparisons can prove very illuminating to explore why industrial health services in Britain did not follow this trajectory. Colin Gordon has recently analysed why America did not establish national health insurance, exploring how private interests shaped the politics of health.[44] In America, private health schemes based within the workplace fuelled the growth of industrial medicine but thwarted ambitions for a nationalised health service. In Britain, the same interest groups – employers, workers, the medical profession and the State – were also engaged in a vigorous debate to determine a course of health policy compatible with their own interests. What resulted was an inversion of the American system, in which industrial medicine was ultimately subordinated to a nationalised health service based within the community. These strikingly divergent responses to the health needs of the working population point to the necessity to study the politics of health and to consider the impact of health care systems in other countries on the development of British health provisions. In Pickstone's account, the State acted as the driving force for productionist medicine, with trade unions and industrial companies playing a supporting role.[45] *The Rise and Fall of the Healthy Factory* will argue that the role played by employee and employer organisations in shaping the course of industrial medicine has been underestimated and will trace the direct and indirect role both groups played.[46] Consequently, this study explores the political history of industrial health (understood as a proactive movement to improve health) which evolved from the contested negotiations between trade unions, employers, the medical profession and the State, as each sought to achieve their objectives through an array of strategies.

The decision to focus on the role played by trade unions and the TUC in the promotion of industrial health might appear imprudent to readers familiar with Paul Weindling's pioneering 1985 edited volume on the history of occupational health, in which Weindling acknowledged that trade unions acquired 'considerable expertise' in the field of workmen's compensation but asserted that 'trade unions have often made pay a priority, and neglected health issues'.[47] This remained a guiding principle in subsequent work: Geoffrey Tweedale, for example, in his study of asbestos-related disease, argued that trade union officials were 'inclined to place jobs above health' and were also prepared to sideline any union members who campaigned against asbestos.[48]

Recent accounts, however, have begun to identify more nuanced trade union responses to industrial illness across different fields, challenging the assumption that trade unions routinely prioritised pay over health and safety. In an article on the compensation of silicosis sufferers, Joseph Melling and Mark Bufton found no evidence to 'support the view that British unions pursued financial awards in preference to campaigns for workplace safety'.[49] Similarly, while acknowledging that a decision taken by the National Union of Mineworkers to approve a subsidy for workers employed in non-approved conditions undermined the Union's health work, Arthur McIvor and Ronald Johnston argued that trade unions played an active role in campaigning for dust control, rehabilitation and compensation.[50] In an analysis of trade union responses to byssinosis undertaken with Sue Bowden, Tweedale found that 'trade unions were deeply concerned about occupational health and had a relatively sophisticated understanding of the medical issues that sometimes pre-empted the experts'. The divergent trade union responses to asbestosis and byssinosis identified by Tweedale appear to be related to the relative power of trade unions in different fields. In their byssinosis study, Bowden and Tweedale acknowledged that unions had concentrated on compensation, but argued that unions only adopted this strategy 'because control of the workplace lay overwhelmingly with the employers and the government'; Johnston and McIvor similarly identified the comparative weakness of trade unionism within workplaces utilising asbestos as a factor which constrained trade union responses to the hazard of asbestos.[51] Reviewing a historiography of occupational illness which has largely sought to attach responsibility and blame to different parties, Melling suggested that the motivations of different actors should be considered in terms of the constraints they faced as much as their perceived interests, an observation that also applies well to the study of industrial health.[52] This study intends to engage with these new perspectives by addressing the role played by the TUC and affiliated unions in the promotion of health within the workplace in relation to the constraints imposed on the TUC by the economic and political context.

Increasing mechanisation, the advent of two world wars, the development of new lighter industries and the relative decline of older, heavier industries and changing social mores meant that the composition of the workforce was constantly in flux during this period. Women and young workers drifted in and out of factory work, though women's participation in the labour force stabilised at just under 30 per cent in the interwar years, while the skilled labourer was increasingly marginalised within industrial production.[53] Trade unions, employers and the State were not oblivious to these trends and all targeted specific groups within the workforce to achieve their own objectives. In the first two decades of the twentieth century, concern was expressed about the place of boys within society as future fathers and defenders of the Empire. Harry Hendrick has analysed how changes in the labour market such as the decline of apprenticeship and skilled labour and the rise of so-called blind alley jobs were perceived to be particularly detrimental to boys' characters. Welfare measures were inaugurated within the factory to curb excesses and provide educational and vocational guidance, forming part of a broader attempt by welfare organisations to inculcate citizenship into boy workers.[54]

The role of women within the factory was perceived as the most problematic, and women factory workers were targeted by statutory and voluntary agencies lest women lose sight of their domestic duties, by employers who viewed female employees as easier to control and by trade unions who viewed women within the workplace as a threat to be either incorporated or expelled. Particular concern was expressed about the health of women factory workers in their dual roles as workers and mothers, reflecting the militaristic goals of the British State in the twentieth century and its obsession with levels of industrial production and the health of the next generation. These anxieties promoted increasing state intervention in maternal health and motherhood.[55] However, it will be argued that concerns about the health of women factory workers transcended their reproductive functions, reflecting new models of vigorous health embracing girls and young women.[56]

In the early decades of the twentieth century attention was increasingly paid not only to the physical harm of industrial work but the psychological damage which could be inflicted on workers engaged in repetitive tasks with little conception of how their work fitted into a broader purpose or those workers who unwittingly found themselves in occupations to which they were unsuited. Both these problems were pursued by the National Institute of Psychology, established in 1921.[57] Such anxieties conflicted with the belief that employment alleviated mental health problems and that unemployment, particularly prominent in the 1920s and 1930s, caused massive psychological maladjustment.[58]

It was not just developments within psychology which shaped the emerging field of industrial health. These new visions of the factory as a site of health improvement reflected broader trends within medicine. Preventive medicine, funded by the State, aspired to curb disease in the mid- to late nineteenth century by improving sanitation. John Simon used his position as Medical Officer to the Privy Council between 1858 and 1871 to undertake investigations into industrial disease and to promote government intervention.[59] In the Edwardian era, preventive medicine sought to break the link between poverty and disease. Urban overcrowding was implicated in the spread of disease and rational town planning advocated; school meals were introduced to help tackle malnutrition.[60] Developments in bacteriology in the 1880s and 1890s focused attention on the role and contingent responsibility of individuals as carriers of disease affecting the health of the community. New knowledge about the causation of disease and the social environments in which it flourished opened up all aspects of human life to medical intervention in order to prevent disease and promote health.[61] 'The first line of defence is a *healthy, well-nourished, and resistant human body*', wrote Sir George Newman in his 1932 book, *The Rise of Preventive Medicine*. 'Nor is the individual, taken at any one moment, the whole of the issues. There is his life-history, his heredity, his family, his domestic life, his personal habits and customs, his home as well as his workshop.'[62] The phrase was lifted verbatim from the final Report of the Health of Munition Workers Committee – chaired by Sir George

Newman – demonstrating the centrality of preventive medicine to the newly emerging field of industrial health.[63]

These concepts of preventive health were slightly ambivalent, premised by a government that until the 1940s was unable to implement a full national health service. While industrial welfarists might seek to suppress visions of the dark hazardous factory and the spectre of industrial illness and accident, in practice new health and welfare initiatives grew in tandem with efforts to regulate minimum standards within the worst factories and secure compensation for workers disabled through accident or disease. Indeed, positive industrial health and welfare could at times be used as a smokescreen by those seeking to camouflage the continuing hazards posed to health by industry, a fact which it has already been suggested might account for the wealth of studies on the hazards posed to health by industry but the relative neglect of industrial health and welfare developments.

In the 1930s and 1940s, interest in the social aspects of disease prevention and health began to coalesce as the discipline of social medicine, finding expression in policy, publications, university chairs, the establishment of a Social Medicine Research Unit under the Medical Research Council and projects such as the Finsbury and Peckham health centres which monitored the physical well-being of nominally healthy adults.[64] Social medicine was seen by some as an extension of occupational medicine, but it took the lessons learned within the factory and applied them outside, prioritising community-based health care services and facilities at the expense of industrial health provisions and personnel.[65]

Structure of the book

Focusing on the two world wars, *Chapter 1* emphasises the importance of the militaristic goals of the British State in the development of industrial health in practice and thought. It explores the direct and indirect mechanisms established by the government to regulate working conditions, studying the role of legislation, committees and agents of the State. *Chapter 2* draws on promotional material created by philanthropic employers and the publications of industrial welfare supervisors and factory inspectors to chart the emergence of the idealised vision of the factory as a site of health improvement. Examined largely from the perspective of trade unions and the TUC, *Chapter 3* explores the politics of industrial health in the interwar years, examining the flourishing interactions between different groups and organisations as they instigated, debated and contested different schemes to improve the health of factory workers. *Chapter 4* charts the emergence of a more differentiated approach to workers in industrial enterprises. While some measures were seen to benefit all employees, young workers and women were believed to have specialised needs which could be met by the emerging army of industrial experts, medical professionals and trade unions. *Chapter 5* explores the place of industrial medicine in plans for a National Health Service, medical training in industrial medicine and efforts after the inauguration of the NHS to introduce a state-run industrial health service. *Chapter 6* traces the decline of the healthy factory, analysing how the terminology used began to shift from industrial health to industrial medicine and then to industrial hygiene. The book concludes by looking ahead to the future path of industrial health in Britain, arguing that its trajectory was intimately related to the relative economic decline of Britain when compared with other economies. The fall of the healthy factory went hand in hand with the decline of the factory and the provision of a comprehensive health care service was seen as neither economically viable nor a priority within a decaying industrial sector which sought to remain competitive in a global market.

1
War and Industrial Health: The Productive Alliance

The State as an agent of political change

'The mere association of war and welfare sounds incongruous', wrote erstwhile Factory Inspector Rose Squire in her 1927 memoir. 'While the State demanded ammunition for the killing of men, it devised means to preserve the health and provide the comforts and recreation for the women who made it.'[1] As the industrial production of the country shifted during the First World War to meet military needs, and thousands of women entered the workforce, the State developed a set of initiatives which would reappraise the functions of the factory and seek to transform the workplace environment.

Histories of twentieth-century Britain have tended to emphasise the emergence and development of the Welfare State, a term which gained currency in the 1940s and has been defined by Derek Fraser as a mode of governance that circumvented the free market to protect the rights of designated groups within society by providing free medical and educational services to all citizens and by offering monetary support through pensions and unemployment benefit to support citizens in times of need.[2] The antecedents of the Welfare State are generally assumed to stretch back into the nineteenth century. Pat Thane, in her account of the emergence of the Welfare State, took 1870 as her starting point, arguing that around this time 'important demands began to arise for the State in Britain to take a permanent...responsibility for the social and economic conditions experienced by its citizens'.[3] Fraser began his analysis earlier, arguing that the industrial revolution marked the emergence of the Welfare State because it created unprecedented social problems which could not be managed within the existing social policy framework.[4] This interpretation supports the thesis advanced by

Jurgen Habermas that the public sphere had collapsed by the twentieth century, leaving the government to arbitrate between different social needs and interests.[5] Consequently, new welfare policies intervened more directly than ever before in the lives of citizens, circumventing the role of public institutions.

Recent revisionist work has challenged the centrality of the State, pointing to the vibrancy of the public sphere evidenced through the actions of voluntary organisations and institutions. Pat Thane's account sought to redress the belief that the voluntary sector was in terminal decline in the face of growing state intervention, stressing its continuing importance.[6] Geoffrey Finlayson made a case for the continuing strength of the voluntary sector throughout the trials of the two World Wars and the lean interwar years, while acknowledging the limitations levelled at the voluntary sector; that it was financially insecure, did not provide uniform nationwide coverage and that it was in some respects undesirable, focusing as it did on the munificence of the charitable rather than the rights of people to services.[7] Habermas' ideas have also come under critical scrutiny and reinterpretation from scholars who stress the continuing vibrancy of voluntary action through the twentieth century, pursued not from a unitary public sphere but an array of organisations and institutions able to represent and embrace pluralistic interests and arguably therefore expand democratic participation.[8]

Focusing largely on the role played by professional organisations and trade unions in shaping the course of industrial health and implementing policy within the scope and limitations afforded in different political, economic and social contexts, *The Rise and Fall of the Healthy Factory* seeks to illustrate the fraught, contested political field in which policy and practice were shaped. This approach has been adopted because a state-centred account of the rise of the healthy factory could only provide a very incomplete account. While a state-run National Health Service was inaugurated, a national industrial health service never came to fruition. It is only possible to fully explore the emergence and decline of this idea by examining the actions of interest groups.

Within this chapter I have taken the decision to study government intervention in factories in both world wars side by side. This approach has been adopted to demonstrate that government involvement in the Second World War to a large extent duplicated the course of action adopted by the government during the previous war. It also reveals that the government's role in the development, advancement and implementation of health within the factory was largely confined to times of war and therefore intrinsically bound up in the militaristic goals of the

British State. This approach is by no means unproblematic; it might suggest that there were no differences in the methods, nature and extent of state intervention between the two wars. This is not the case and the differences will be fully analysed and discussed in the conclusion to this chapter. The adoption of this approach might also suggest to readers that the interwar years were a period of stagnation as regards the promotion of industrial health. Quite the contrary: as the next two chapters will demonstrate, efforts to promote the ideal of the healthy factory flourished in the interwar years. However, the main agents in this movement were professional groups, trade unions and employers: government involvement was minimal.

During both wars, the government extended a hitherto undreamt of degree of control over private industrial production, the employment of workers and civilian life.[9] Shortly after the First World War, Charles Baker proclaimed in a study undertaken for the Carnegie Endowment for International Peace that a great extension of government control and regulation had taken place during the war over public utilities, health and safety and the welfare of wage earners. It might be accepted, Baker wrote, that 'government operation and control of industry in the past four years has not merely driven the last nail into the coffin containing the defunct *laissez faire* theory of government; it has dumped that coffin without benefit of clergy into the grave...and has heaped high the earth over it'.[10] However, both wars can be characterised by not only the intensity of state intervention but also the short-lived nature of that intervention.

This chapter will argue that the State effected change within the workplace through a variety of different mechanisms. In both wars, new legislation enabled agents of the State to modify working environments through regulations. Committees were established under the auspices of government departments, which drew upon the evidence and expertise of professionals to create recommendations that then informed the regulation of industry. New groups of workers, claiming professional knowledge, sought to govern conditions within and outside the workplace to achieve a variety of objectives. Direct state control over industrial production expanded dramatically during both wars. By 1943 the Ministry of Supply had become the largest employer in the country, with 300,000 workers employed in 42 Royal Ordnance factories. Unprecedented levels of intervention also extended over private industry to control recruitment and wages and prevent industrial action.[11] By drawing attention to the interaction of military objectives and welfare initiatives, this chapter supports David Edgerton's recent account which critiques the centrality of the Welfare State, 'rather than the welfare *and* warfare states' in the historiography of modern Britain. It also reflects the development of 'productionist medicine', a phrase coined by John Pickstone to describe a phase of medicine instigated by the State which intended to improve the health of the nation in order to meet imperialistic and industrial needs in the first half of the twentieth century.[12]

The pioneering work undertaken by the Health of the Munition Workers Committee (hereafter the HMWC) during the First World War will be contrasted with the work of the Industrial Health Research Board (IHRB), Industrial Health Advisory Committee (IHAC) and Factory and Welfare Advisory Board established during the Second World War. It will demonstrate that despite the hopes expressed in the aftermath of the First World War, and despite the continuation of research work by the Industrial Fatigue and Industrial Health Research Boards, many of the lessons supposedly learnt between 1914 and 1918 were subsequently discarded. The waning of state interest in industrial health in the aftermath of both world wars demonstrates that the primary objective had not been to establish a permanent system to safeguard and improve the health of workers but to maintain production levels for the duration of the war.

Politics: legislation and committees

The changes that took place within the workplace during the First World War were only partially brought about by direct legislation. Initially, under section 150 of the 1901 Factory Act, the Home Secretary issued orders relaxing the provisions regarding hours of employment for munitions factories. The annual reports of the Factory Inspectorate suggest that many companies extended their hours without applying to the Home Secretary, believing that the Factory Acts were 'in abeyance'.[13] Hours of work were subsequently restricted in many industries. The 1916 Police Factories etc. (Miscellaneous Provisions) Act enabled the Home Office to issue orders for health and welfare provisions: regulation was thus facilitated but not directed through legislation. Welfare orders made under the 1916 Act included an order requiring drinking water in every factory or workshop employing 25 people or more, an order for the provision of seats for female workers in munitions factories and a first aid and ambulance order which required an ambulance room in factories that fell within the scope of the order and employed 500 or more people.[14]

Much of the business of analysing and resolving the problems posed to health by the workplace was left in the hands of committees. The Ministry of Munitions appointed the HMWC in 1915 'to consider and advise on questions of industrial fatigue, hours of labour, and other matters affecting the personal health and physical efficiency of workers in munitions factories and workshops'.[15] In its lifetime, the Committee met 39 times, interviewing employers, workers and factory inspectors in London, Birmingham, Sheffield, Newcastle, Glasgow, Manchester, Coventry and Woolwich. Committee members visited factories to interview employers, foremen and workers and to ascertain at first hand the conditions under which munition work was being carried out. The Committee produced 21 memoranda and two reports, of which over 210,000 were sold or distributed. Factory Inspector Adelaide Anderson contended that many of these memoranda, which began to be published in the autumn of 1915, circulated not just among employers but reached 'a wide, general, reading and thinking public'.[16]

The inauguration of the Committee and its scientific approach was not welcomed by all commentators. An editorial from *The Times* responded critically to the infiltration of the factory by experts, asserting that the issue was one of common sense:

> The problem is really practical, not scientific; but the Committee consists mainly of medical men and government officials...That people get tired, that they do not work well when they are tired, and that the remedy is rest, are truisms familiar to the whole of mankind since the remotest ages. Nor is any inquiry needed to prove that eight hours a day are less fatiguing than twelve and six days a week than seven. People know when they are tired and when they are fresh better than any doctor can tell them.[17]

Some organisations were established explicitly to address the influx of women into the workplace. The Home Office and the Board of Trade instituted the Women's Employment Committee to consider issues arising out of the mass substitution of women from men, such as housing, transit facilities, canteen provision and arrangements for recreation. Local advisory committees were then established, coordinated by a Central Advisory Committee: these were concerned mainly with welfare outside of the factory and coordinated voluntary efforts. In 1916, the Home Office and Board of Trade issued a series of pamphlets on the substitution of women in industry for enlisted men which offered advice on health and welfare arrangements for women workers.

In 1916, the Ministry of Munitions established a Welfare Section under Seebohm Rowntree to undertake the executive work arising from the Committee's recommendations, promoting the health of the large body of workers for which the Minister was responsible.[18] This section superseded the Women's Employment Committee and the Central Advisory Committee. The Department was at first concerned to secure improved conditions of health and welfare within the controlled factories by encouraging the appointment of welfare supervisors. The 1918 *Annual Report of the Chief Inspector of Factories and Workshops* attributed improvements in working conditions in many staple industries to the extension of government control over such industries during the War and the accompanying visits of welfare officers of the Health and Welfare section of the Ministry of Munitions.[19]

The Department was reorganised in 1917 under Dr Edgar Collis, a member of the HMWC, and was made responsible for all matters concerning the health and welfare of munition workers in national and controlled factories. The sphere of the Department expanded to incorporate conditions of life of munition workers outside the factory, including the inspection of large numbers of temporary hostels provided for them. In 1918, Rose Squire was seconded from the Factory Inspectorate and was appointed the Director of Women's Welfare. Towards the end of the War, the Department started to address the various issues posed by maternity and work.

Some committees were established to tackle specific issues. A Canteen Committee of the Central Board of Control was established in June 1915: this Committee was responsible for controlling alcohol, leading to protest from some workers' unions. It was also involved in the provision of canteens, seen as a constructive measure to counter alcoholism. Efforts were made to tackle the housing of workers displaced from their homes by the Hostels Department of the Ministry of Munitions and, from 1917, the Billeting Board.

The outbreak of war led the government to extend control over many staple industries, to initiate legislation which enabled agents of the State to introduce improvements in working conditions and to inaugurate committees and departments which were instructed to investigate ways of improving workers' health and output. These findings and suggestions then became the basis for many welfare orders and shifts in practice. Thus, while the initial impetus to improve health and welfare within the workplace may have rested with the government, they subsequently entrusted this task to other organisations such as the Young Women's Christian Association. 'The change in the mode of industrial legislation may be summed up as a tendency to move from the politician towards the expert', wrote Dorothea Proud in her account of welfare work.[20] 'A definite and considerable advance towards national control of the means of welfare of factory workers has been achieved during the great War, by means of legislation as well as administration.'[21]

At the start of the Second World War, the mechanisms established during the First World War to promote industrial health had largely disappeared. As no centralised stimulation or coordination of the activities concerned with industrial welfare existed, it was suggested that the Ministry of Labour and National Service should undertake this task as the Ministry of Munitions had done during the First World War. In June 1940, the Factory Department and the administration of the Factory Acts were transferred from the Home Secretary to the new Minister of Labour and National Service, Ernest Bevin, recently seconded from his post as General Secretary of the Transport and General Workers' Union by Churchill. 'I realise all that was done in the last war and was associated with a good deal of it', wrote Bevin for a speech given in 1943. 'I am afraid in the time between the two wars a good deal of it was lost; it drifted and there was a loss of impetus at a later stage'.[22] Seeking to make good the ground lost since the First World War, Bevin established new organisations while reviving existing bodies such as the IHRB which had its origins in the Industrial Fatigue Research Board, itself born from the ashes of the HMWC. In 1940, the IHRB published an emergency report which detailed its previous work and invited industry to bring forward new problems. Receiving little feedback, the IHRB undertook other research of a non-industrial character, including the clothing of military personnel and the ventilation of air-raid shelters.[23] The IHRB was reconstituted in 1942 when its terms of reference were extended to encompass psychological effects and the prevention or amelioration of illness caused by exposure to particular hazards at work.

The Minister convened the Factory and Welfare Advisory Board in the same month and established a Factory and Welfare Department, which was intended to work in close cooperation with the IHRB, local authorities, trade unions, employer organisations and voluntary organisations. New local officers, known as welfare officers, dealt with welfare matters outside the factory, such as board and lodging, transit, day nurseries, arrangements for shopping facilities and the reception of transferred workers, under the Ministry and in conjunction with local representatives of other government departments. The Central Consultative Council of Voluntary Organisations coordinated welfare work outside the factory. This extra-mural work became increasingly important as growing numbers of married women with domestic responsibilities were employed on a part-time basis in order to meet labour requirements.

The 1940 Factories (Medical and Welfare Services) Order provided similar powers to the 1916 Police Factories etc. (Miscellaneous Provisions) Act.[24] It enabled the Ministry of Labour and National Service to make arrangements for medical supervision, nursing and first aid services and welfare supervision of people employed in factories manufacturing or repairing munitions of war or producing articles required for the manufacture and repair of munitions or undertaking any work on behalf of the Crown. The Factories (Canteens) Order issued in 1940 empowered the Factory Inspectorate to compel a manager who employed more than 250 workers on government work to institute canteen facilities.[25] Levels of lighting within wartime factories were subject to the Factories (Standards of Lighting) Regulations of 1941 which recommended a substantially higher standard of lighting than previously suggested as a legal minimum.[26]

Despite the establishment of new advisory organisations, a report issued by the Select Committee on National Expenditure on the health and welfare of women in war factories published in 1942 noted that while 'many valuable reports dealing with various aspects of health and welfare problems have from time to time emanated from departmental, technical and advisory bodies... little attention seems to have been paid to many of the constructive suggestions contained in these documents'. The Report argued that neglect and an absence of coordination characterised government policy, blaming the multiplicity of departments and overlapping interests involved. The Ministry of Health was not empowered to deal with health problems within industry, while the Ministry of Labour lacked manpower and organisation; these problems would persist in the post-war context.[27] In March 1943, the Ministry of Labour and National Service acted upon the recommendations of the Select Committee and appointed an Industrial Health Advisory Committee (hereafter IHAC). Simultaneously, the Ministry announced that a three-day conference on industrial health had been arranged for April 1943 to emphasise the importance of industrial health and to elicit suggestions for its promotion. Dermatitis, radiological and ophthalmological advisory panels were established to address significant areas of concern.

To some, the establishment of the IHAC just added to the multiplicity of organisations and hindered coordination of medical services. In the House of Commons Bevin was forced to defend his decision to establish the IHAC; the purpose of the IHRB was to undertake research, he explained, while the IHAC was created to advise Bevin on the technical and scientific problems arising out of the current administration of the department.[28] In both World Wars, ministers set up committees and boards which enabled the government to bring industrial experts and representatives of employers and workers into consultation, although the number of organisations established could at times hinder rather than help the implementation of a coordinated policy. This pattern would re-emerge in the post-war context as the government established a profusion of committees and sub-committees to investigate the future development of industrial health services.

War and the development of industrial medicine

The decision taken by the Ministry of Munitions during the First World War to issue an order requiring the provision of first aid and ambulance room facilities within factories employing more than 500 persons increased medical facilities and personnel within industry, though quantifiable evidence as to exact numbers of doctors and nurses exercising medical supervision within factories are difficult to obtain. Statistics relating to the provision of first aid and ambulance room facilities in 1919 reveal that 85 out of a total of 290 ambulance rooms were staffed by a fully trained nurse, while a further 43 were run by a partially trained nurse.[29] To put these figures in context, these 290 ambulance rooms were spread among the 135,454 factories and 145,737 workshops listed in 1919.[30] Factories engaged in the production of TNT and employing over 2,000 workers were also required to provide medical supervision by employing a full-time medical officer, though the number of doctors so employed does not appear to be listed. Deborah Thom, who has examined the role of doctors employed in factories using TNT, argued that their role was administrative, not curative, and was 'to remove workers who were particularly likely to die, as long as there were not too many of them'.[31] Notably, the doctors were employed not by the government but by the factory management.[32]

In the Second World War, the requirements of the Factories (Medical and Welfare Services) Order of 1940 provided a near identical stimulus to the growth of medical personnel and facilities within factories, albeit on a far larger scale. By the end of 1942, approximately 850 works medical officers were employed within industry, of which just under 19 per cent were employed full time.[33] By August 1943, 8,385 nurses were employed within factories, of which around half were state registered.[34] By the end of 1944, 180 full-time doctors were providing supervision in 275 factories, while a further 890 gave part-time service to 1,320 factories.[35] Again, it is worth comparing these figures for medical supervision against the total number of factories operating in 1944: 228,776.[36] As in the First World War, doctors within industry were employed and paid by the factory management, despite the concerns expressed by the TUC, which wanted to see doctors employed by the State and chosen by a joint committee of workers and employers. Within some of the Royal Ordnance factories, rehabilitation, chiropody, dental and ante-natal services were established.[37] Despite paper shortages, the *British Journal of Industrial Medicine* was launched in 1944; a testimony to the 'enormous development of industrial problems with a medical aspect' which industrial medicine had to manage and resolve.[38]

As the spotlight turned upon medical provisions within factories during the Second World War, it seemed an opportune moment for members of the medical profession regarded as experts within the field of industrial health to outline their vision of what an industrial medical service should look like and how it could be developed after the War. John Bridge, Senior Medical Factory Inspector of Factories from 1926 to 1942, expressed the view that little specialist knowledge was required by a doctor wishing to practise as an industrial medical officer and that it would therefore be misleading to regard industrial medicine as a distinct medical specialism, though medical students would benefit from a month's experience working in a factory ambulance room or clinic.[39]

His successor, Edward Merewether, had a more ambitious plan for the post-war industrial medical service. While the wartime industrial health services provided only partial provision, Merewether wrote, they 'fill a gap in the Health Services of this country which it is inconceivable can be allowed to disappear when the immediate emergencies are past'.[40] He envisaged the post-war development of the service and urged the provision of further training to counter the shortage of specialist personnel. For Merewether, industrial health was 'a composite subject comprising health, safety and welfare of man in his working environment', and an industrial health service team should therefore comprise of 'doctors, engineers, chemists, physicists, psychologists, welfare and personnel experts, dieticians, and other specialists'.[41] This conception of the industrial health services would resurface in post-war debates about the shape a proposed industrial medical service would take. Even during the war, the reports of the medical factory inspectors questioned how growth would be sustained given other demands on medical manpower and the shortage of doctors and nurses with special training and qualifications in industrial health. H. A. Waldron asserted that the number of industrial medical officers decreased substantially in the post-war period, slumping from the 180 employed full-time in

1944 to a paltry 51 by 1951, 'as though industrialists had quickly forgotten the benefits they had derived from them in more stringent times'. This argument of employer indifference was premised on a misreading of a table: in fact, the number of full-time industrial medical officers remained virtually the same.[42] Nevertheless, the same argument could convincingly be made if one turns from the number of doctors employed within industry to the number of nurses, which had halved from over 8,000 in 1943 to 4,000 by 1951.[43]

New approaches to industrial health adopted during the two World Wars

The medical facilities and personnel introduced into the workplace covered only a small proportion of wartime factories; efforts to maintain the health of the industrial workforce in both wars was based more on a holistic model of health which placed emphasis on preventive health care and the relationship between the individual and their environment. The First World War witnessed an unprecedented interest in manipulating the general workplace environment to ameliorate the health of workers. Much of the factory legislation existing prior to the outbreak of the war had been implemented primarily to protect women and children by restricting the number of hours they could work and prohibiting their employment in more dangerous trades where workers were at risk of developing industrial diseases.[44] In contrast, the reports of the HMWC suggest that it interpreted its remit as being concerned with health and preventive medicine, discarding paternalistic and philanthropic approaches to the welfare of workers and adopting in their place a scientific approach which stressed the impact that the workplace environment and working hours could have physiologically on the worker. To secure efficiency and maximum output, the HMWC maintained that the machine must be subservient to the man, a view they termed the 'physiological basis of labour'.[45] Physical health was therefore the fundamental basis or key ingredient in successful industrial production, and contrary to received wisdom on industrial illness, the HMWC asserted that 'it is the conditions of employment rather than its character which undermine the physical strength and endurance of the worker...the dominant evil is not accidents or poisoning or specific disease'. The main hazards to health, the HMWC argued, were the long working hours which produced fatigue and stress and did not allow the worker time to rest and enjoy recreation. 'It is not the work', it claimed, 'but the continuity of the work which kills'.[46] The HMWC's interest in fatigue, working hours and productivity were revived during the Second World War.

This reorientation within the field of industrial health should be contextualised within the emergence of preventive medicine, with its emphasis on health rather than illness and its desire to study the whole man within his community. The HMWC was chaired by Sir George Newman, the Chief Medical Officer of the Board of Education, and an advocate of a state medical service and preventive medicine. This excerpt from a report of the HMWC relates this emerging discipline to the field of industrial health and specifically to the capacity of individuals to work effectively:

> The science and art of Medicine is not restricted to the diagnosis and cure of disease...It is, in fact, the science and art of Health, of how man can learn to live a healthy life at the top of his capacity of body and mind...able to work to the highest power, able to rest to the fullest, growing in strength and in the *joie de vivre*...there is thus a psychological aspect of preventative medicine hitherto greatly neglected. Nor is the individual, taken at any one moment, the whole of the issue. There is his life history, his heredity, his family, his domestic life, his personal habits and customs, his home as well as his workplace...The subject of industrial efficiency in relation to health and fatigue is thus in large degree one of preventative medicine, a question of physiology and psychology, of sociology and industrial administration.[47]

During the Second World War, the psychological approach to industrial health, already in evidence during the First World War, eclipsed the physiological. In practice, interest in the same environmental factors persisted, but their effects were understood through a psychological rubric. Government committees also showed an interest in the relationship between health issues and the factory environment, exploring non-industrial health problems which nevertheless affected health and productivity. 'There is a distinction between industrial health, in the sense of healthy conditions and practices in the industrial environment', announced Bevin in a press release issued in March 1943, 'and the health of the individual worker in which industrial health is only one although an important factor.'[48]

These themes were elaborated upon at the industrial health conference organised by Bevin in April 1943 at which Dr A. J. Amor, the deputy Chief Medical Officer of the Ministry of Supply, expressed the view that the health of factory workers could not be reduced to 'pills and plasters'. He described the environment at one Royal Ordnance Factory, 'set in a lovely garden, with hanging baskets of flowers'. Here, Amor explained, a beauty parlour had been established for the female employees, who worked on detonators. They did their work better, Amor claimed, 'because of their sense of well-being and looking well'.[49] Illustrating, however, just how indiscernible the line was between health, well-being and industrial illness, Amor made a causal link between the installation of the beauty parlour and a 60 per cent decline in the incidence of dermatitis at this particular factory, though offering no further explanation as to how the beauty parlour had achieved this. Dermatitis was the most common industrial disease experienced during the Second World War, with cases rising from 2,952 in 1939 to 4,744 in 1940. However, as the example quoted above indicates, both its causation and prevention were frequently linked to worker behaviour and general well-being rather than to industrial materials or processes. Senior Medical Inspector John Bridge asserted that most cases of this industrial disease could be prevented by keeping skin clean through regular washing, suggesting that the answer lay in greater personal hygiene. An opposing view was expressed by the TUC's Social Insurance Secretary J. L. Smyth, who stressed the causative role of the materials used in industry, arguing at a meeting of the IHAC that all new substances should be examined and vouched for by the State before they were used in industry.[50] Notified cases of dermatitis continued to rise, numbering 8,802 in 1942 and reached a peak of 8,962 in 1943.[51]

Illness and the factory environment

The example of the beauty parlour and its supposed effects on the rate of dermatitis was a rather extreme variant of a more general belief frequently expressed by government bodies and officials during the Second World War that incidental factors were believed to affect health and productivity in the workplace. 'The causes of specific occupational diseases are necessarily emphasised in this report', wrote Bridge in his chapter for the 1939 Report, 'but it must not be thought that factors affecting the general health of the worker are not equally matters of concern.'[52] The IHAC spent much of its time attempting to curb the spread of disorders which were not specifically linked to the factory environment but which nevertheless threatened to curb productivity, such as colds. Colds, a memorandum noted, were 'one of the principle causes of ill-health resulting in absence from work and loss of production'. The IHAC were asked to explore whether workers suffering from colds should be asked to stay home to avoid spreading the disease, or whether other steps could be taken to minimise the risk of infection within the workplace. Possible solutions explored by the IHAC either sought to sanitise the factory environment or targeted the worker directly and included asking sufferers to wear a mask at work, the use of disinfecting sprays or UV light to sterilise the atmosphere, experimentation with inoculation to prevent the spread of colds among industrial workers and the extent to which adequate ventilation could 'dilute' the dosage of infection. Health advice also stressed how workers' resilience could be increased through workplace conditions and provisions such as nutritious canteen meals. Production rather than health was clearly the primary concern. A note from the Medical Research Council recommended that workers with colds should report to work, partly to deter the slacker from absenteeism and partly because the sick worker might well infect others in shops and cinemas if not in the factory. The IHAC decided that the efficacy of inoculation was unproven and, noting that no effective treatment existed for colds, suggested there was no point advising workers to seek medical assistance.[53]

Investigations undertaken by the IHAC into treatment methods for rheumatism and the potential use of radiology to screen for tuberculosis and other abnormalities of the lung reflected the new interest shown by the government in the interactions between the workplace, treatment and illness. While these diseases were related to the working environment, they were not recognised as industrial illnesses as such. Rheumatism received attention because, like colds, it was a common cause of lost working time, even if not regarded as an 'industrial' disease; Ministry of Supply figures revealed that 10 to 12 per cent of male and 7 to 8 per cent of female employee sick leave was ascribed to rheumatism. The threat posed by rheumatism to industrial productivity was taken so seriously that the IHAC explored the possibility of adopting a treatment procedure not recognised by general medical practice.[54] The Ministry of Labour also considered the acquisition of a mass radiology unit for use within factories both to detect tuberculosis and to pick up any abnormalities of the lung caused by the inhalation of dust. It was not anticipated the screening for dust-related abnormalities would lead to treatment; the Ministry suggested it would help identify hazardous working conditions which could then be ameliorated though the introduction of preventive measures to reduce further occurrences.[55]

Approaches to dermatitis and colds had both emphasised the importance of maintaining general standards of health and cleanliness, and general health education was vigorously promoted to industrial workers from several different directions during the Second World War. The Ministry of Labour issued an illustrated pocket-sized pamphlet entitled *Fighting Fit in the Factory*, designed to convey information to factory workers as to how they could maintain their health.[56] The pamphlet contained generalised health hints which stressed the value of exercise, fresh air, sunlight, plenty of sleep, a healthy diet with vitamin C and cod liver oil. Other tips on posture and suitable footwear sought to ensure the comfort of employees and women were warned of the hazards of wearing their old dance shoes for work. Advice literature aimed at women munition workers during the First World War had similarly emphasised the importance of hygiene, fresh air, exercise, a healthy diet and sensible clothes and footwear.[57] The Central Council of Recreative Physical Training and the Football Association established the 'Fitness for Service' scheme which provided men with the opportunity to participate in games and physical training at weekends and in the evening. This scheme was extended in November 1940 to enable women to take part after the Ministry of Labour and National Service decided to reimburse the Council for the expense incurred by men taking part in the programme who were engaged in work of national importance. Another voluntary organisation, the Central Council for Health Education, established a factory sub-committee which arranged the distribution of health advice pamphlets including 'Rest and Relaxation', 'Care of the Feet' and 'Be Kind to Your Stomach', of which over 60,000 of each were distributed to factories in the month of January 1943 alone.[58]

The Central Council was also involved in publicising the risks of venereal disease; initially the Ministry of Labour had placed an embargo on the distribution of a pamphlet which discussed venereal disease, *Health of the Worker*, to factories on the grounds that workers might see this as class discrimination. Dr Robert Sutherland, Medical Adviser to the Council, argued that factories were targeted not because of the specific health risks posed by industrial work, or because the working class were believed to be in particular need of health education, but because 'industry offers an opportunity for reaching a larger and more representative section of the adult population than can be reached elsewhere'.[59] In many respects, this approach characterised health interventions in factories during the Second World War. Workers were targeted at factories not so much because of the particular hazards industrial work posed to health, but because of the sheer number of people gathered together in one place (a fact which in itself facilitated the spread of contagious disease), and because their ill-health would impede industrial production. Subsequently, the Ministry of Labour gave permission for distribution of the *Health of the Worker* pamphlet and for meetings to be held in factories: in the year between April 1942 and May 1943, meetings on venereal disease were arranged at 300 factories with an approximate total audience of 115,000.[60] Simultaneously, the British Social Hygiene Council, supported by the Ministry of Health, also sought to educate people about the hazards of venereal disease through the medium of the workplace and contacted the TUC to enlist its support for a campaign among 14 to 18-year-olds, claiming that 'the only certain place where they were to be found collected together was the factory'. Rather adroitly, the TUC's General Secretary Vincent Tewson suggested that the National Youth Council would be a more suitable medium to pursue this objective, though the TUC later acceded to a request to facilitate lectures and film shows arranged by the Council at halls in factories and other undertakings.[61]

Feeding the factory 'army'

The development of canteen provision within factories during both World Wars, premised on the grounds that good nutrition was a fundamental requirement for good health, reflects how medical developments and interventions remained at the margins of the project to transform the factory into a site of health production and maintain levels of productivity. While the Factory Inspectorate cited a few exemplary cases of canteen and mess room provision prior to the First World War, such as an initiative to provide dining centres in London for the use of working girls in industrial areas which were run by voluntary social workers,[62] they more commonly bemoaned inadequate provisions. The canteen movement of the First World War stemmed from the initiatives of the Canteen Committee of the Central Board of Control, with the assistance of the YWCA and YMCA. Before the War, there were around 100 regular factory canteens. By the end of the war there were around 1,000 canteens either in operation or under construction. In addition to or sometimes as an alternative to canteens, some factories would send round trolleys with drinks and snacks, or have coffee and snack stands. In many factories, long journeys to work and badly timed rest breaks made it difficult for workers to take adequate meals, while in others canteen facilities were not available at night. The location of the canteen within the factory, its cleanliness, layout, lighting, décor and ability to provide hot, affordable, varied and nutritious meals rapidly to large numbers of workers were all considered by the HMWC.[63]

Many writers viewed nutritious food as the basis of health, one of the most effective means of preventing workers from succumbing to illness or industrial poisoning and, consequently, a stimulant to industrial output for both male and female workers. 'Output in regard to quality, amount and speed is largely dependent upon the food of the workers', asserted T. H. Agnew in a report on the health and physical condition of male workers undertaken for the HMWC.[64] 'If "an army fights on its stomach" is it not also true that a factory works on it?' argued Factory Inspector Rose Squire.[65]

Feeding workers re-emerged as a pre-eminent concern during the Second World War. Indeed, if anything the demand for canteens was greater than during the previous war. Heavy bomb damage to workers' homes stimulated demand for canteens alongside rationing, shift systems, shorter meal times and increasing employment of married women. Returns from factory inspectors early in 1940 revealed that canteens had been established in 64 per cent of the works engaged mainly in the making of munitions and employing over 500. In parts of southern England, the figure was as high as 90 per cent.[66] Factory canteen advisers were appointed following the Factories (Canteens) Order of 1940 to advise employers on issues such as layout, equipment, staffing, menu-planning, cooking and the nutritional problems of large-scale cooking.

Canteen provision rose from 5,695 in December 1941 to 8,481 by December 1942, by which stage 98 per cent of the factories subject to the Factories (Canteens) Order 1940 had canteens in service or in active preparation.[67] By December 1944, 11,630 factories had canteens and a further 180 were in preparation.[68] The Factories (Canteens) Order 1943 superseded the 1940 Order and extended the scope of provisions to include factories catering for civilian needs as well as the manufacture of munitions.[69] Around 80 per cent of factories reportedly employed a manager to run the canteen, reflecting the growing specialisation of canteen provision. Those in charge of canteens were urged to ensure the nutritional value of meals, to provide snacks for those who did not want large meals, to consider the challenges of providing food for night shift workers and to entice young workers to eat sensibly by offering discounts on hot meals. Work carried out in Scotland found that young people tended to grab tea and rolls for breakfast and lacked vegetables in their diets. One young worker responded, 'I've never eaten vegetables, and I'm not starting now.'[70]

As James Vernon has argued, the objectives underpinning the extension of factory canteens transcended the provision of food to hungry workers. Canteens were intended to help educate the workforce about sensible food choices, to encourage civilised behaviour by placing workers in a civilised environment and to foster a social atmosphere within the factory.[71] Attention was thus paid not only to the nutritional value of the food on offer in factory canteens but to the appeal of the food and the canteen to the worker. Standards in canteens varied, however, and rationing limited the variety that could be achieved in menus. The 1942 Report of the Inspectorate complained that many canteens relied too heavily on boiled cabbage and did not seek other ways to provide vitamin C, while the standard of the tea was a further frequent cause of complaint. Other canteens failed to appeal to workers because no attention had been paid to décor, though the shortage of building materials meant that many canteens had to be established in makeshift premises. Growing attention was paid to the difficulties of providing food to small firms. Efforts to establish joint canteens between groups of factories met with little success, while transporting food from British restaurants to small factory mess rooms proved logistically difficult and impracticable during night shifts, though the Inspectorate did note a large increase in the number of canteens provided without any legal obligation by small factories: of the 10,577 factories that had been established by the end of 1943, 5,704 were situated within factories employing fewer than 250 people. In 1943 the Factory Inspectorate noted that when undertaking routine visits to small works, inspectors frequently received complaints regarding the absence of canteens from workers 'who cannot understand why a distinction should be drawn between large and small firms, and who are growing to feel that they too are entitled to a canteen'. [72]

Adapting the factory environment to the worker

One example of how self-proclaimed experts sought to alter the factory environment to improve productivity during the First World War can be seen in the attention devoted to lighting standards. In its final Report, the HMWC discussed the need for consistent lighting which should be shaded to prevent glare and placed so as to not cast shadows on work. Given the resurgence of night work the standards of artificial lighting assumed a new prominence. The Committee stressed that, where possible, natural lighting was preferable to artificial lighting and urged employers to keep windows clean and to paint walls and ceilings light colours to reflect light.[73]

After the cessation of the First World War, government interest in lighting standards in factories rapidly declined. However, the construction of new factory buildings with no natural lighting, the adaptation of buildings originally designed for other purposes and the installation of fixed blackout screens in many factories for safety purposes during the Second World War all exacerbated lighting problems. The Factory Inspectorate noted that complaints regarding inadequate ventilation were frequently received from workers employed in permanently blacked out workshops which subsequently were found to be adequately ventilated, and surmised that the root cause was psychological. So widespread did this phenomenon become that the term 'blackout anaemia' was coined to describe the condition. Factory inspectors found that poor standards of artificial lighting aggravated the problems of a lack of daylight. In 1941, the Ministry of Labour and National Service issued new regulations for standards of lighting within factories in line with the recommendations issued by a departmental committee report.[74] Despite the new regulations, standards in many factories were found to inhibit production while even good lighting installations performed inadequately because they were not cleaned regularly or provided with shades to prevent glare. In 1942, the Factory Inspectorate urged employers that 'at a time when there are no fresh sources of manpower to draw upon, any means such as this of assisting production becomes of added importance', a request which suggested that health be prioritised because of the labour shortage.[75] Figures from one division covering nearly 600 factories found that in 18 per cent improvements were necessary and in a further 42 per cent improvements were necessary but could be obtained without a great deal of expenditure on materials and labour. The most common defect was found to be glare due to unscreened lamps, a fault that had been noted back in the First World War.[76]

The lighting problems experienced within wartime factories were illustrated in the Mass Observation report of a war factory. The observer described her experiences working in the assembly shop which had been constructed during the war and had no provision for natural lighting. Permanent assembly workers who complained to the observer about the lack of daylight in the shop stated that while the artificial light did not impair their vision, it did affect their state of mind. 'They say this light's very good for your eyes, but it can't be like daylight', one worker was reported as saying. 'It's not right to keep people shut in all day long like this.' An employee who had previously worked in the assembly department when it was located within an old country house compared her new workplace unfavourably: 'It's all clammy down here, like being in prison', she stated. 'We sat there watching them putting it up, and we knew that once it was finished we'd never see the sun any more.'[77] These workers reflected the shift in emphasis which occurred between the two wars; what had once been mainly characterised as a physiological problem in which impaired vision compromised production had been transformed into a psychological problem whereby poor lighting standards affected workers' state of mind and consequently inhibited their productivity.

From a physiological to a psychological understanding of industrial fatigue

The issue of hours of work came under discussion both because of its impact upon the health and welfare of workers, and its potential to influence levels of industrial output. In its reports and memoranda, the HMWC claimed that excessive hours were counterproductive, leading to a decline in output, while sickness rates and lost time escalated.

> The personal health and physical efficiency of the munition worker, as of all industrial workers, are measurable by two standards – first, that of fatigue, weariness and exhaustion of the healthy physical faculties and functions of the individual; and secondly, that of disease.[78]

It was not simply long hours, but an absence of or badly timed rest breaks, the undertaking of monotonous tasks and working night shifts which predisposed workers to industrial fatigue. Studies carried out by the Committee into the health of operatives in different factories found marked levels of fatigue among workers employed for long hours: 15.5 per cent of women employed in one factory for 77 hours weekly, for example.[79]

Although it was frequently argued that there was no conflict of interest between the interests of workers and managers, and consequently the interests of the State which relied on high levels of industrial output to meet wartime needs and requirements, intervention in what had been seen as a matter for industrial relations between trade unions and employers under the guise of industrial welfare was a politically fraught matter. These concerns were addressed by the HMWC in the introduction to its *Interim Report*, which noted that 'a suspicion has grown up amongst workers that any device for increasing output will be used for the profit of the employer rather than for the increased health and comfort of the workers'.[80] However, just a few pages later H. M. Vernon wrote:

> I understand that the object of the Committee is in many ways similar to that of the managers of munition works, and is to ascertain the hours of employment most likely to produce a maximum output over a period of months, or maybe even years...If health and physical efficiency are maintained they would raise no *a priori* objections to any given number of hours, however long.[81]

One researcher for the Committee, commenting on the levels of fatigue among girls working night shift in a cartridge factory, suggested that he did not find the level of fatigue which verged on sleep that problematic as long as output was maintained. He observed that the girls tended to fall asleep the moment their machine broke down and marvelled that while they appeared to be 'continually on the verge of sleep', output was maintained. He concluded that the workers had learnt to 'automatise' their movements and were able to perform their task even while half-asleep.[82] While night shifts were believed to be too taxing for married women who would often forgo sleep in the day to fulfil their domestic tasks, it was argued that many girls preferred night shifts as they gained more time for recreation and shopping, and higher wages as well. Lilian Barker, the Government Inspector of Munition Workers' Welfare at Woolwich, justified the long hours of workers with reference to the welfare provisions available to them:

> The twelve hour day was not hurting them one little scrap...This long day was possible because the conditions of work at Woolwich were good...They got good money and they spent it largely on food. They had better boots and better macintoshes and they were far better, healthier and stronger girls than they had been.[83]

Wage systems also came under consideration by the HMWC, which analysed the differential impact of piece rate systems versus time rates upon output and nervous strain. Wage systems, it argued, should be understood by the workers and should encourage workers to increase output but not to the extent of endangering their life, while offering a fixed minimum income to prevent the less adept workers becoming anxious. The HMWC claimed that 'the health, *i.e.*, the absence of sickness, physical and mental, and efficiency of workers is influenced by their earnings, and that output, which has been closely investigated as an indication of fatigue, may be influenced by the wage system in force'.[84]

The assertion that good working conditions and higher wages paid by war industries benefited workers' health has been amplified by Jay Winter, who asserted that workers were able to supplement their diets which in balance led to an improvement of health.[85] However, efforts to improve workers' health by improving the environment of the factory and installing canteens was compromised by the negligible extent of medical supervision available to workers and the failure of the HMWC to persuade many employers to adopt shorter working hours. Myra Baillie's research on the health of women munition workers in Clydeside concluded that the model new welfare provisions introduced in factories had little impact in mitigating the debilitating effects of 12-hour shifts which continued to be worked and did little to reduce the risk of industrial illnesses such as TNT poisoning.[86] Even within the Committee's own reports, the deterioration in health of many employees was noted. A follow-up inquiry undertaken by Janet Campbell in 1917 on the health of women munition workers depicted a significant decline in the health of many employees from her earlier investigations of 1915 and 1916.[87] Of the 1,183 women examined, over 40 per cent showed signs of fatigue or ill-health. Campbell described many of the women employed on 12-hour shifts making 6-inch shells as 'completely exhausted and unfit for recreation or work'; 'many had the appearance of rapid loss of weight, and those who did not know their weight could remember that their clothing had to be definitely taken in to make it fit the present waist'.[88]

While the research undertaken by the HMWC was informed by a desire to increase production, its findings could be used to bolster the demands of workers by providing a scientific justification for shorter hours. One article published in *The Times* proclaimed that 'many industrial disputes could be settled at once by scientific investigation … recent scientific work of the very first order has suggested that the man who asks for shorter hours is very frequently right, and the employer who opposes him is just as frequently wrong'. The article welcomed the creation of the Industrial Fatigue Research Board which was established to investigate how hours of labour and methods of work related to industrial fatigue.[89]

One might imagine that by the time of the Second World War, employers would have accepted that longer hours did not necessarily result in greater productivity. However, this does not appear to have been the case. The 1941 Annual Report issued by the Factory Inspectorate reasserted its belief that reasonable hours of work and good working conditions were essential to obtain maximum production and success-fully prosecute the war. However, it noted that 'a good deal of time has to be spent in convincing even some government departments that this is so'.[90] The working hours of women and young persons, capped under the 1937 Factory Act to 48 hours a week and 44 hours for young persons under 16, could be extended under section 150 of the Act up to 58 hours and 48 hours respectively for workers employed making munitions and other wartime requirements if a general emergency order was obtained. When France surrendered in May 1940, regulated working hours fell by the wayside and working hours of 'extravagant proportions' were instituted.[91] After July, hours of employment of protected persons were brought under control again, though the Inspectorate disapprovingly noted that the tendency towards long hours of work and seven week-days persisted for male employees who were not covered by the 1937 Factory Act. By the end of 1940, 5,493 factories held permissions to operate under the General Emergency Order. Total weekly hours varied between 48 and 54 in 1,585 factories; 55 or 56 in a further 1,027 facto-ries and over 56 to 60 in 2,428 factories.[92] By 1941, General Emergency Orders increased from 5,493 to 9,129. Individual orders were also in operation, covering about 10,000 factories which could employ women and young persons for more than 48 hours or on a system of day and night shifts.[93]

Year after year, the Inspectorate provided examples of how productiv-ity had been maintained or even increased and absenteeism reduced by the adoption of shorter hours. In 1941, a factory was described at which women had worked 56 or 57 hours for six days, and now worked a 52-hour, five and half day week. Absenteeism, which had been asso-ciated with shopping and other domestic problems, consequently declined.[94] Long hours were also implicated in the spiralling rise in accidents which befell women workers over the course of the Second World War. Between 1938 and 1942, the number of reportable accidents occurring to women rose by 389 per cent from 14,626 to 71,244, while male employees experienced a 51 per cent increase in the accident rate, which rose from 134,752 to 203,865. The disproportionate rise in accident rates among women was also partly attributed to the grow-ing number of women taking on dangerous jobs in heavy trades, the greater likelihood their clothes or hair would catch in machinery and the acceleration of production. The shortage of maintenance staff and of supervisory staff contributed to the accident rate, while the Factory Inspectorate asserted that 'the ailing worker is more liable to accident than one in good health' and argued that adequate medical supervision would do much to reduce accidents.[95]

In January 1941, the Liverpool Union of Girls' Clubs started a rest break scheme to provide fortnight breaks for women and girl workers in the Liverpool area. The scheme was instituted in response to the number of workers on Merseyside suffering from strain because of a combination of physiological and psychological factors: air raids, long hours and blackout conditions. The first rest break hostel was established in April in Abergele and was designed as a preventive service for the 'flagging worker' who was suffering from accumulated fatigue and was liable to breakdown if she did not get a rest.[96] The Ministry of Labour and National Service subsequently issued a memorandum which described the scheme and a Rest Breaks National Advisory Committee was formed which decided to establish hostels near all important industrial areas.[97] The following year, the Ministry issued a leaflet entitled *Making the Best of Holidays in 1942*, which gave advice on how to provide attractions for the home-staying holiday maker.[98]

If fatigue was in part psychological, the monotony of undertaking repetitive tasks could be as hazardous as long working hours. The Ministry's *Fighting Fit* pamphlet sought to jolly along the worker engaged in monotonous work by encouraging him or her to think of the contribution they were making to the war effort. 'Every hour, nay every minute, of dull, hard work put in is helping the cause', the leaflet explained, suggesting that 'the lucky man or woman can keep his mind busy, while his hands are at work, on something that interests him'.[99] However, while in the First World War it had been assumed that women workers were 'naturally' able to switch off and undertake monotonous work without suffering any detrimental psychological effects, driven by a patriotic urge to do their bit for the war effort, it was recognised during the Second World War that this might not be the case. William Elger, General Secretary of the Scottish TUC, described conditions at industrial camps situated in Inverness-shire and Ross-shire in his capacity as Chairman of the Scottish Consultative Committee for Voluntary Organisations. The most disheartening aspect of the visit, Elger wrote, was being informed that 'the men showed no interest in the national importance of the work they were doing – that it was just another job with better money than usual'.[100]

'For an intelligent worker...to turn out on an automatic machine a small component without any appreciation of where that particular jobs fits into the whole scheme of production of the finished article, seems not only unkind, but lacking in vision', the Report of the Factory

Inspectorate noted in 1941.[101] Some factories held exhibitions to demonstrate how each individual task contributed to the final product in an effort to awaken an interest in workers in their work. Even this appears to have met with limited success. The machine shop girls described by the Mass Observation observer displayed little interest in their work, but a great deal of interest in the passing of time. Most were conscripts who had not chosen the work and whose fantasies revolved around leaving the factory when the war was over. The Report described an exhibition that had been held of the finished sets and, while this had provoked interest, the observer felt 'their interest was personal rather than mechanical; that they did not care how or why the part fitted on where it did, but that they had a strong feeling of being personally complimented on their work'.[102]

Much of the work undertaken in the Second World War may have been unavoidably and inherently dull, but efforts were made to alleviate the tedium experienced while at work. The introduction of the *Music While you Work* programme by the BBC was reported in 1941 to have been of assistance.[103] The Entertainments National Service and the Council for the Encouragement of Music and Arts provided concerts for workers during lunch hours, in evenings and during weekends in local halls or within the workplace to help alleviate boredom,[104] while workers stranded at more remote locations could sometimes expect visits from mobile cinemas.

Meeting individual needs: from welfare supervision to personnel management

The visions of industrial health and welfare promulgated in the publications of the HMWC were to be largely carried out and monitored by a relatively new type of worker, the welfare supervisor. Before the outbreak of the war, welfare supervision within factories had been an isolated phenomenon, largely directed towards female employees and received favourably by the lady factory inspectors who claimed that women supervisors improved the cleanliness, discipline and moral tone of the workplace and workers, overseeing bathroom and mess room arrangements. In some instances, such welfare supervisors were also cited as managing other provisions such as uniforms, seating for workers within the factory and holiday homes.[105]

The recommendations of the HMWC that welfare supervision would greatly benefit the health of workers and the welfare orders subsequently issued mandating the appointment of a welfare supervisor in many factories led to an upsurge in demand. Intensive training courses were established by the government to train suitable women to fill these vacancies.[106] Dr Collis estimated that approximately 1,000 women welfare workers were appointed and 400 for boys, although he complained that the low salaries frequently offered for this work made it difficult to attract high-calibre applicants. 'It has been no uncommon thing', he wrote, 'to hear some stupid act of tyranny quoted as an example of "welfare".'[107] In her study of women welfare supervisors, Angela Wollacott noted that women who entered the field of welfare supervision came from an array of middle-class backgrounds; some had been employed as social workers, others had backgrounds in health care or academic study, while some were promoted from the factory floor.[108] The class difference between workers and welfare supervisors was a frequent source of conflict. Trade unions also expressed disquiet that welfare supervisors were recruited by and paid for by employers and not the State, a pattern which was repeated with the employment of industrial medical officers and nurses.

Although welfare supervision was introduced on a national scale during the First World War largely to cope with the issues and problems believed to be posed by the influx of female labour into traditionally male industries, this era also witnessed the emergence of welfare supervision for boys and men. The Boys' Welfare Society was founded in 1918, establishing its periodical, the *Boys' and Industrial Welfare Journal*, in 1918.[109] Chaired by Sir William Beardmore, of W. Beardmore & Co, and directed by the Reverend Robert Hyde, the 21 members of the Society's Council included 15 company directors, three MPS, Sir Robert Baden-Powell and two trade unionists, both of whom represented the Amalgamated Society of Engineers. This employer-dominated movement viewed the ideal welfare supervisor as:

> a man of upright character, patient, tactful; a man of good temper, capable of acting sympathetically as a friend and adviser; above all, he must be a man who can command the respect and entire confidence of the boys. He should organize their games, recreations, sports, and attend them, and should visit the boys in their homes occasionally, taking a part also in their private lives.[110]

Another organisation, the Central Association of Welfare Workers (Industrial) had 15 local branches and 700 members by May 1919. This group aimed to promote and develop industrial welfare work, maintain standards of work through training and qualifications, arrange conferences and lectures to exchange and discuss ideas and to collect data.

During the First World War, welfare supervisors began to call for professional organisation, advancement and university training for supervisors. A conference was held in 1917 at the Home Office between representatives of universities and other educational authorities on the training and selection of welfare workers. In a speech given to the London School of Economics, Professor Urwick addressed the need for specialisation and decent wages, describing welfare supervision as a skilled job which was 'beyond the scope of anyone who has not made a careful study of the conditions of it ... It requires essentially detachment as well as knowledge'.[111] However, it is difficult to see how skilled many of the supervisors recruited during the course of the First World War were, given that the courses provided by universities to train supervisors were only of a few weeks' duration; the HMWC recommended that future courses should last at last a year.

While welfare supervisors were once more in heavy demand within industry during the Second World War, the profession was increasingly termed 'personnel management', reflecting a shift in focus from physiology to psychology. 'It has been increasingly appreciated', claimed the 1941 Report of the Factory Inspectorate, 'that the welfare of employees implied attention not only to their physical comfort but to their mental and psychological make-up also', which was best met through the institution of good 'personnel management'.[112] This change in emphasis had in part been driven by welfare supervisors themselves who sought to professionalise their role after the First World War, establishing new professional associations which stressed the need for training to enter the field.[113]

At the start of the Second World War, around 1,500 welfare supervisors or personnel managers were employed, a similar figure to the number employed within industry by the end of the previous war. The Ministry of Labour instituted courses of three months' duration at four universities to counteract the shortage of qualified personnel. Candidates for courses were selected by committees which included members of the TUC, the universities involved, employer organisations and the Ministry of Labour in an effort to overcome the criticisms that personnel managers were middle-class ladies with no understanding of trade union politics. Some 2,000 people were interviewed, of which 800 were selected for the training, consisting of lectures for two months and one month of practical work. Later on in the war, part-time courses were arranged for those already employed in personnel management within industry. There were eight such courses, and between 700 and 800 people attended.[114]

As in the First World War, welfare supervisors or personnel managers formed the largest group of professionals overseeing the health and welfare of factory employees. By 1941, in larger factories employing over 500, about 90 per cent had some systematic form of supervision.[115] Of the 4,774 factories employing more than 250 people and undertaking work which could bring the workplace within the scope of the Factories (Medical and Welfare Services) Order, 3,395 employed between them 5,478 welfare supervisors or personnel managers.[116]

It may have been redefined, and there may have been more of an attempt made to ensure politically neutral candidates through the selection process, but personnel management in the Second World War was by no means an unqualified success. Between 68 and 69 per cent of factories were described by factory inspectors as having satisfactory personnel relationships; in 5 per cent personnel relationships were considered definitely unsatisfactory and in the remainder they were considered indifferent. This was attributed in part to inadequate training: the standard peacetime course lasted two years. In other instances, personnel management had not been a success because welfare had been compartmentalised and personnel managers had been given no say in the engagement, placement and dismissal of workers, no control over the health and safety of the workplace or no input into the organisation of the works canteen.[117]

Conclusion

By analysing developments which occurred during the First and Second World Wars, this chapter has sought to demonstrate how methods of intervention, research interests and hopes expressed for future development correlated across both conflicts. In many respects, the State's response to industrial health in the Second World War simply replayed its actions of 20 years previously.

This parity of experience is remarkable given the proclamations issued during the course of the First World War and in the immediate aftermath concerning the important lessons supposedly learnt from scientific investigations into factory conditions and working hours. 'Among all the changes which the Great War has brought in its train,' wrote the Minister of Munitions David Lloyd George, 'none is more significant, and none more likely to have a lasting effect than the revolution in the structure of British industry.'[118] The opportunity proffered by the War,

Lloyd George argued, must not be allowed to slip and the results of war-time initiatives he argued should be 'of permanent and enduring value, to the workers, to the nation, and to mankind at large'.[119] The Minister's sentiments were echoed by the Factory Inspectorate, which wrote in its 1918 Report that the experience gained during the war

> has shown that an increase in the ordinary hours of work does not necessarily mean an increase in production, that the more the comfort and welfare of the worker is studied and provided for the greater is his output, and that attention given to health and the prevention of accidents means greater efficiency.

'The result will,' it asserted, 'to a large extent, be permanent.'[120]

Even *The Times*, which had been so critical of the adoption of a scientific approach to workers' health and output in an editorial responding to the initiation of the HMWC in 1918,[121] enthusiastically embraced the movement by the end of the war, indicating that the climate of opinion had undergone a fundamental shift during the course of the war:

> Readers...are aware that the movement [industrial medicine]...has had our support for some time, ever since, indeed, the studies carried out on the health of munition workers revealed the great urgency of a scientific examination and organisation of labour power. Industrial medicine has been called the new political economy. Perhaps it is rather a new kind of police force, designed to protect industry from fatigue, bad ventilation, accidents, and disease, by which the worker is deprived of his health, the employer of his output, and the State of its strength.[122]

Given this apparent enthusiasm for industrial health from all quarters, it is curious that the government's response to industrial health during the Second World War played out like an unimaginative sequel to a blockbuster film. This raises three issues for further exploration. First, while not exaggerating the extent of innovation during the Second World War, one must acknowledge the differences in approach and in the sheer scale of the measures adopted. Secondly, the unavoidable fact that much ground had been lost and had to be recouped needs to be explained. This demands an interrogation of developments during the interwar years; it requires us to study what happened to the idea of the healthy factory and the mechanisms established to promote it. We must also turn attention on the complex political field of action and interaction on the subject of industrial health in the interwar years, reconsidering the role of the State as a major player within the field. These fields of enquiry form the subjects of the next two chapters. Finally, if the promising state of affairs at the end of the First World War evaporated with little to show for itself, it did not augur well for developments after the Second World War.

An examination of the issues explored by committees, the Factory Inspectorate and research organisations demonstrate that to a great extent, the same problems dominated both wars. Seeking to maintain industrial output, the issue of worker fatigue and its relation to working hours was perhaps the major issue. Other subjects related to worker fatigue and output recurred, namely how fatigue might be ameliorated and health maintained through diet and factory canteens, and how workers could be encouraged to persist in monotonous jobs. Both wars saw the government rely on new personnel to supervise the health, welfare and productivity of factory workers, namely welfare supervisors, nurses and doctors. Yet a closer exploration reveals important distinctions in theory and scope. In the First World War, a physiological approach had predominated: investigations undertaken by the HMWC explored the physical impact of hours of work and working conditions on the health and output of employees and sought to adjust working routines to minimise physical fatigue, though the psychological impact of working hours was increasingly recognised. While attention was paid to maintaining the physical health of the employee during the Second World War by promoting health education, diet, exercise and rest, increasingly the responses of workers to types of work, working conditions and hours was understood in psychological rather than physiological terms.

Another crucial difference was in the sheer scope of intervention. By the end of the First World War, around 1,400 welfare supervisors were employed to oversee the welfare of employees; the comparable figure at the end of the Second World War was 5,478. Similarly, though on a smaller scale, the number of medical personnel employed within factories was far higher during the Second World War. By the end of 1944, 180 full-time doctors were employed and a further 1,320 on a part-time basis to provide medical supervision to a total of 1,595 factories.[123] While the employment of industrial medical officers during the First World War was noted, and required in factories employing over 2,000 persons which utilised TNT, the numbers so employed are not listed. More definite figures can be found for nursing: by August 1943, a total of 8,385 nurses were employed within industry, of which around half were state registered. In 1919, a total of 128 nurses were employed in factory ambulance rooms, and a third of these were only partially trained. These diverging standards of provision cannot be explained by an increase in the number of workplaces covered within the remit of the Factory Inspectorate between the two wars. Indeed, registered workplaces declined from the figure of 281,191 listed in 1919 to 228,776 given in the 1944 report.[124]

At the cessation of hostilities in 1944, Senior Medical Factory Inspector Edward Merewether looked with satisfaction at the promising state of the field of industrial medicine. A variety of organisations, Merewether noted, had made statements in support of industrial health. The Royal College of Physicians had issued the Report of its Social and Preventive Medicine Committee on Industrial Medicine in January 1945 while the TUC had issued a statement and conference resolution urging the development of industrial medicine. Meanwhile, research and teaching into the subject were expanding. Durham, Glasgow and Manchester Universities were all in the process of establishing departments of industrial health with assistance from the Nuffield Foundation, a body which had also made provisions for fellowships in industrial health for those who wish to qualify for senior teaching and research posts in the subject. The Society of Apothecaries was instituting a diploma in industrial health while the Medical Research Council had established a Research Unit at Cardiff to study pneumoconiosis in coal miners. The Medical Research Council's Department of Industrial Health Research was undertaking research and the Royal College of Nursing was offering training to nurses in industrial work. Brief courses on industrial health had been established by the Universities of Bristol, Leeds, Sheffield, Manchester, the London School of Hygiene and the Birmingham Accident Hospital. In the field of industrial first aid, the St John Ambulance, British Red Cross Society and St Andrew's Ambulance Associations were collaborating to raise the standard of knowledge and practice of first aid in industry.

Promising local initiatives were also in the offing: the Leeds Joint Council on Industrial Medicine, comprising of employers, employees and medical professionals, coordinated activities in the sphere of industrial medicine and similar councils had been established in Derby, Burton-on-Trent, Leicester, Nottingham and Stratford. Merewether hoped these initiatives would help expand the pool of medical personnel qualified to work in the field of industrial health. War conditions, he wrote, had pointed to the need to eliminate occupational disease. They had also, however, restricted the means of meeting it, causing a shortage of trained personnel, of equipment and suitable buildings.[125]

Merewether's enthusiasm for industrial health was shared by A. W. Garett, the Chief Inspector of Factories, who applauded the growth of works committees and the greater voice consequently being given to workers which he saw as heralding a 'new era' in which workers would exercise effective supervision over the working conditions. 'Can we develop towards the idea that a factory is more than a place to work in?' asked Garrett.

> A factory is a place where goods are made and they must be made economically and profitably or we will get neither goods nor wages; but also we must consider that a factory is a place where people spend one-third of their working lives and must, therefore, be so run that the well-being of the worker is secured.

Garrett concluded with a word of warning. 'All that I have said above was said almost in so many words at the end of the last war and was forgotten in the intervening years leaving the inspectorate with little support in these matters', he wrote. 'Can we see that it does not occur again?'[126]

While militaristic needs had propelled the State to promote industrial medicine, expanding the number of medical personnel within factories and inaugurating new research bodies, a number of potential stumbling blocks had emerged during the Second World War that threatened to impede progress. On the one hand, there appeared to be no consensus as to what industrial medicine was exactly and therefore what personnel would be required. While a new journal for industrial medicine was launched, retiring Medical Factory Inspector John Bridge had asserted that there was no need for a specialism of industrial medicine as such;[127] his successor Edward Merewether promoted the idea of an industrial health as opposed to an industrial medical service, in which engineers, chemists, personnel officers and dieticians would work alongside doctors.[128]

The role of the State in any future industrial medical service was also hotly disputed. During both wars, employers had been responsible for hiring and paying industrial medical officers, setting a precedent which displeased many trade unionists and other groups such as the Socialist Medical Association which believed that responsibility to the management compromised the impartiality of industrial medical officers. This topic was fiercely contested in the post-war era when proposals for a national industrial health service were afoot. Staffing a nationwide industrial health service also seemed increasingly problematic, and the wartime demands on medical personnel would be replaced in the aftermath of the War with the staffing needs of the new National Health Service.

The final major stumbling block to the post-war development of an industrial health service was the dawning realisation that the term 'factory' encompassed a diverse array of establishments, and that while it might be comparatively straightforward to provide good working conditions and medical supervision within large factories, this would not be the case for small establishments. At the industrial health conference organised by Bevin in April 1943, this was made apparent. 'A "factory" may mean anything', noted a reviewer for the journal *Public Health*, 'from a wretched workshop up to a "place of industry".' He concluded the term 'factory' to be largely meaningless.[129] During the first meeting of the IHAC, TUC representative Ann Loughlin 'stressed the need for realising that the average factory was not a well-equipped large modern factory, but small and old-fashioned. Illness amongst industrial workers was too often the result of bad working conditions in factories of this kind.'[130] The apparent neglect of this issue before the Second World War can be partially explained by exploring how the ideal of the large, modernist healthy factory took hold of the popular imagination in the interwar years.

2
The Rise of the Healthy Factory

Documenting his journey across England in 1933, J. B. Priestley recorded his response to the factories sited on the outskirts of London. 'Years in the West Riding have fixed for ever my idea of what a proper factory looks like', he noted: 'a grim blackened rectangle with a tall chimney at one corner.'

> These decorative little buildings, all glass and concrete and chromium plate, seem to my barbaric mind to be merely playing at being factories. You could go up to any one of the charming little fellows, I feel, and safely order an ice-cream or select a few picture postcards. But, as for industry, real industry with double entry and bills of lading, I cannot believe them to be capable of it... Actually, I know, they are tangible evidence, most cunningly arranged to take the eye, to prove the new industries have moved south... At night they look as exciting as Blackpool. But while these new industries look so much prettier than the old... they also look far less substantial. Potato crisps, scent, toothpastes, bathing costumes, fire extinguishers; those are the concerns behind these pleasing little facades; and they seem to belong to an industry of little luxury trades... But if we could get a living from them, what a pleasanter country this would be, like a permanent exhibition ground, all glass and chromium plate and nice painted signs and coloured lights.[1]

Priestley's recollections illustrate the countervailing trends which characterised British industry in the interwar years. On the one hand, a utopian optimism shaped post-war reconstruction, as efforts were made to establish new model factories which would improve the lives of those who worked within them. On the other hand, depression gripped large swathes of the economy, limiting the resources for improvement and causing unemployment levels to rise. The post-war boom in Britain was much weaker than the global trend and was interrupted by the General Strike of 1926; consequently the world Depression had a less catastrophic impact upon the economy than countries which had fared better in the immediate post-war years. Heavy industries such as steel, iron and shipbuilding, largely based in the north of the country, were heavily reliant on exports and overseas investment and consequently went into decline.[2] New industries such as the manufacture of cars, radios, light aeroplanes, artificial silk and the film industry flourished. Older premises in cities began to be replaced by new model factories, many of which were situated in suburbs in the South and London district rather than in city centres.

Priestley's response to these new model factories of the interwar years conveyed an air of disbelief; were these highly publicised glamorous facades merely a chimera which distracting attention from the true character of British industry in this era? This chapter investigates the extent to which industrial workplaces began to cater for the health needs of their employees in the interwar years. It situates these developments within the economic and cultural climate of this era, exploring how two overlapping types of model factory – the homely garden factory of arcadia and the modernist industrial crystal palace – were constituted in advertising in the interwar years. In so doing, it demonstrates how representations of health helped draw together production and consumption between the wars and points to the vast gulf which existed between the small number of iconic model factories and the majority of industrial workplaces in Britain.

The factory in twentieth-century Britain

In the nineteenth century, critics lambasted the factory as a social evil which undermined the health and morals of the working classes. The mechanisation of work and erosion of skilled labour as mass production expanded in the early years of the twentieth century prompted new anxieties, which we can trace through the growing metaphorical use of the term 'factory' to describe, at best, large-scale, unthinking, mind-numbing, non-individualised production within education, popular culture and the film 'industry'. Writing in 1944, Theodor Adorno and Max Horkheimer attacked the triumph of business over art within a culture industry which manufactured cultural commodities designed for categorised groups rather than reflexive individuals. The culture industry, they wrote, occupied 'men's senses from the time they leave the factory in the evening to the time they clock in again the next morning with matter that bears the impress of the labour process they themselves have sought to sustain throughout the day'.[3]

The *Oxford English Dictionary* noted the first use of the term 'factory farm' in 1890 in a footnote of Alfred Marshall's *Principles of Economics*, in which 'the recognised principles of factory management would be applied, machinery would be specialised and economised, waste of material would be avoided, by-products would be utilised, and above all the best skill and managing power would be employed'.[4] In the 1920s, industrial farming concepts began to be applied to farms in America.[5] In Britain also, a growing interest was displayed in the potential benefits of transferring factory methods to farming. Speaking at a 1926 joint session of the Agricultural and Economics sections of the British Association, Lord Bledisloe expressed his belief that better business organisation, the development of labour-saving machinery and factory farming methods could enhance British farming.[6] While these early advocates stressed the enhanced efficiency and output which could be reaped by the application of factory techniques to farming practice, the term would subsequently come to denote inhumane farming methods in which animals were reared in cramped conditions. If techniques developed within the factory could be applied to the raising of livestock, they could also be utilised when contemplating their demise. Daniel Pick asserts that, from the 1860s, 'technology, factory production and calculated death were coming together in new ways' which would transform not only the slaughterhouse but methods of mass murder.[7] Zygmunt Bauman argued that Auschwitz was 'a mundane extension of the modern factory', its end product death. One defendant at the Nuremburg trials disingenuously sought to defend his role in the Final Solution by arguing that he had simply been 'a cog in a relatively low position of a great machine': a phrase that had come to typify the malaise of workers employed in large-scale industrial enterprises.[8]

This might well suggest that factories developed along the same bleak course envisaged by the Victorian critics of the factory system cited in the introduction. Yet in the 1920s and 1930s the adjectives used to describe factories were poles apart from these images of horror: factories were airy, sunbathed, homely, modern, healthy and even beautiful. Similarly, if we explore not the things which were metaphorically described as factories but the metaphors used to describe factories, we find that readers in the 1920s and 1930s were encouraged to imagine factories as palaces or utopias. Indeed, modernist architects and women's groups seized upon ideas and practices formulated within the factory – rationalisation, scientific management, mass production and standardisation – as liberating forces which could resolve housing problems and ease the burden of domestic labour within the home.[9]

We can begin to explore how this transformation occurred if we turn our attention to William Morris, best known as a utopian thinker, radical Socialist Conservative and advocate of the arts and crafts movement, or as his Wikipedia entry reductively asserts, 'as a designer of wallpaper and patterned fabrics'.[10] Morris depicted the factory system as a product of capitalism which fostered class oppression to line the pockets of the rich, forcing people to undertake degrading, inhumane work to produce useless goods which pandered to the frivolous tastes of the upper classes. In a series of articles published in *Justice* in 1884, he described how workers were subordinated to the machine and slaved for long hours in squalid, overcrowded factories.[11]

Morris's solution, however, was not to abolish the factory. Instead, he envisaged a socialist factory of the future, imbued with art and beauty, which would enrich the lives of workers, reinvigorating the production process with craft and skill. If the wealth and land of the factory owner was redistributed, Morris could see no reason 'why the highest and most intellectual art, pictures, sculpture and the like should not adorn a true palace of industry'. Nor would the factory be situated in a grim urban sprawl; it would 'stand amidst gardens as beautiful...as those of Alcinous', which would be tended by the factory workers as a means of open-air recreation.[12] Morris's idealised factory, working for the common good rather than profit, would not pollute its surroundings. Furthermore, in addition to the space set aside for industrial production, it would contain other facilities and spaces: a dining hall, a library, a school. This would enable workers to pursue their education and hold entertainments.

Within the factory itself, mechanisation would not be disposed of; instead, Morris inverted the received wisdom that man had been subjugated by the machine. Once the profit motive had been discarded and factories simply produced goods that people needed, there would be no need to stockpile merchandise. Mechanisation would thus liberate workers, relieving them of perfunctory chores, enabling them to undertake more artistic tasks and complete their factory work within a four-hour day. Morris's vision of the ideal factory drew on a tradition of utopian writing and strikingly paralleled psychiatrist W. A. F. Browne's reinterpretation of the asylum as utopia, published nearly 50 years before in a lecture entitled 'What Asylums Ought to Be'.[13] Browne argued that a well-designed and constructed asylum could improve the health and well-being of its residents, emphasising the importance of an attractive rural location, gardens, and recreational and educative facilities. In Browne's vision, the ideal asylum 'appears a hive of industry ... it is as if you had entered the precincts of some vast emporium of manufacture; labour is divided, so that it may be easy and well performed, and so apportioned, that it may suit the tastes and powers of each labourer.'[14]

The arts and crafts movement, of which Morris was a leading figure, developed in late nineteenth-century Britain in response to what its proponents perceived as the devaluation of creative design and the work of the craftsman. Reformers sought to transform working conditions, reasserting the value of craftsmanship and well-designed products. In practice, however, Morris found it difficult to reconcile his arts and crafts ideals with business imperatives, dismayed by the 'contradiction between his ideal of a democratic art and the "idle privileged classes" who formed his patrons'.[15] Charles Harvey and Jon Press have argued that Morris was heavily influenced by the writings of John Ruskin but was selective in his implementation of Ruskin's prescriptions and that, consequently, the employment he offered 'fell short' of these ideals.[16] Thus, while Morris focused on the production of high-quality products, most of his workers had little scope to exercise their creativity in the production process where a division of labour operated.

In the twentieth century, Morris's vision of the ideal factory was replicated by other thinkers who also wished to humanise the system of factory production by countering meaningless, degrading work processes and reinvigorating the interior and exterior of the factory. The parity of thought on the ideal factory displayed by Morris and later would-be reformers extended beyond this, however. By simultaneously embracing modernity, with the promise of liberation through technology, while advocating pastoralism and the arts and crafts, Morris typified a dialectic that would persist in thinking about the ideal factory in the interwar period. These later representations of the ideal factory were however largely denuded of Morris's socialist vision, seeking to calm rather than fan the flames of class conflict and political strife.

It is this pastoralism that differentiates idealised British factories and their American counterparts. In David Nye's account, the modern American factory was an emblem of technological progress, revered as a sublime industrial object.[17] In Britain, technological awe went hand in hand with arcadian pastoralism; ideal British factories were envisaged as a reinvention of the pastoral idyll, homely factories in gardens, situated in garden cities. In his analysis of literary representations, Raymond

Williams noted that while the Industrial Revolution transformed the city and the country, 'English attitudes to the country, and to ideas of rural life, persisted with extraordinary power... even in the twentieth century, in an urban and industrial land, forms of the old ideas and experiences still remarkably persist'.[18] W. Warde Fowler's 1893 memorial for eighteenth-century naturalist Gilbert White claimed that 'the spread of the factory system, and the consequent growth of huge towns, has strengthened rather than weakened this love of all things rural. We pine for pure air, for the sight of growing grass.'[19] Despite owing their fortunes to the factory system of production, Martin Weiner argued that the emergent industrial elite subscribed to an aristocratic culture which elevated rural life and the countryside over urbanisation. In his view, British industrialisation was not so much a revolution as a compromise in which a sustained process of 'psychological and intellectual de-industrialization' contributed to British industrial decline.[20] Yet Weiner's interpretation does not fully account for the developments in factory design in the interwar years. As David Matless has demonstrated, modernism pervaded the interwar landscape preservation movement. Preservationists sought to realise an ordered vision of the English landscape which embraced modern technologies and remedied inefficient and chaotic layouts.[21] Similarly, Bill Luckin has revealed how electricity was portrayed as a means of revitalising rural life, capable of breathing new life into farming and rural crafts.[22] As these studies illustrate, many believed that technology and mechanisation, far from being inimical to the preservation of English countryside, could help preserve rural lifestyles from the onslaught of chaotic urban and industrial squalor.

The image of the grim Victorian factory looms so large in the popular imagination that remarkably little has been written about the factory in twentieth-century Britain. Yet throughout the twentieth century efforts were made to radically redefine what the functions of a factory were and what a factory should look like. The layout of textile mills in the eighteenth and nineteenth centuries had primarily been determined by practical considerations; reliance upon natural lighting, for example, restricted the width of mills. By the early twentieth century, industrial engineers sought to enhance the efficiency of industrial production by experimenting with methods for handling materials and directing work flows through the factory. Increasingly, as engineers' attention turned to how space could be used to manage the human element within the system of production, they advocated open plans designs which facilitated surveillance and minimised unnecessary worker movement.[23] With the shift from water to steam power, manufacturers were able to use heavier machinery, thus necessitating larger buildings.[24] The introduction of steel and concrete facilitated these developments as they provided greater strength to buildings, allowing larger windows to be installed and reducing the number of columns necessary for support, giving rise to a new modernist vision of the factory: the industrial crystal palace.[25]

In their study of the industrial archaeology of Britain in the twentieth century, Michael Stratton and Barrie Trinder challenged the assumption that factories were functional buildings lacking architectural merit. Instead, they argued, factories built in the twentieth century were constructed with a range of objectives in mind, whether it was the need to incorporate new technology, a desire to make an architectural statement, to create a more user-friendly workspace or simply to construct a building in keeping with its surroundings.[26] Similarly, in her study of the factories designed by architectural firm Wallis, Gilbert and Partners in the interwar years, Joan Skinner contended that factories were not just functional buildings designed solely to contain the process of manufacture.[27] Or rather, that the perceived function of factories had extended to incorporate protecting the health of its workers and enhancing their productivity through attention to colour schemes, lighting, heating and ventilation. Skinner referred to factories constructed by Wallis, Gilbert and Partners in the period 1927–1935 as 'fancy' factories, 'commanding advertisements in a competitive world',[28] designed to challenge the perceived status of industry and of industrial employees, to be noticed rather than overlooked. These factories, she argued, were 'fanciful, dressed out in glorious colours, catching the eye and the imagination, romantic, lifting the spirit, freeing the mind from toil and grind. Making the goods, attracting the buyers, improving the lot of the nation – and celebrating it.'[29]

Following the pioneering investigations of the Health of Munition Workers Committee (HMWC), commentators questioned whether factory work and the factory could be transformed to improve the lives of factory workers. In the immediate aftermath of the First World War, working conditions within factories were very diverse. While many new premises were being constructed, often single-storey buildings which allowed plenty of daylight into the building and with provisions for welfare amenities, the Factory Inspectorate noted the difficulty experienced by occupiers of older buildings in just meeting the minimum legal requirements. At the height of the Depression, the Factory Inspectorate noted a curtailment of voluntary welfare provisions: as unemployment rose, employers no longer had to compete to attract workers.[30]

The proportion of factories to workshops shifted decisively in favour of factories with growing mechanisation. Factories, distinguished by a supply of electricity, surpassed workshops for the first time in 1924, numbering 142,494 as opposed to 133,729.[31] This trend continued to accelerate throughout the interwar years. In 1930, there were 154,102 factories to 103,371 workshops,[32] while by 1937, 169,277 factories were listed compared to 71,106 workshops.[33]

The interwar health movement

In the interwar years, a number of individuals, organisations and institutions began to stress the promotion of health rather than the treatment of illness, snowballing into an international health movement.[34] Speaking at a health education conference in 1930, Lord Riddell typified this interwar approach to health:

> More social and political movements were aiming directly or indirectly at securing better health, or the means of improving it. Demands for better wages, better housing, better education and the readjustment of our economic system were in reality part and parcel of the race for health and long life. Too much attention was apt to be paid to the abnormal, to the neglect of the normal ... It was essential to create what was called a 'health sense'.[35]

This positive message was fuelled by aspirations for a better life in post-war Britain but also constrained by financial limitations and haunted by the detrimental impact on physical and mental health of unemployment and poverty.[36] The new Ministry of Health appointed Sir George Newman, who had chaired the HMWC, as its first Chief Medical Officer in 1919. Newman adopted a holistic approach to health, seeking to advance personal health care as well as ameliorate the environments in which people worked and lived.[37] Nevertheless, Charles Webster argued, official reports produced by the Ministry of Health downplayed the prevalence of malnutrition and morbidity among the population, recognising that the root cause of the situation was largely economic and that the Ministry lacked the resources to remedy the situation.[38] The interwar health movement can also be characterised by its conflicting impulses towards and against modernity. Voluntary organisations such as the National Council for Mental Hygiene, the People's League of Health, the New Health Society and the Health Education Council preached a message of preventive health care and promoted a state of health.[39] They often stressed the benefits to health of close commune with nature, urging people to take regular exercise, eat a healthy diet,[40] have access to fresh air and sunshine,[41] and suggested that new buildings such as schools and houses could be designed to help fulfil these goals. These ideas found expression in the establishment of health centres which sought to uncover states of sub-health within people and remedy incipient health problems.[42] Moreover proposals for a state medical service, which proliferated in the interwar years, tended to emphasise a preventive rather than a curative approach. The Socialist Medical Association produced a number of publications in the 1930s on the subject of a national preventive health service which were inspired by health care schemes inaugurated by other countries such as the USSR, publicised by Henry Sigerist's 1937 volume.[43] While a growing body of research testifies to the vigour and international character of this interwar health movement, little if any attention appears to have been paid to the fact that plans to reform the factory and consequently improve the health of the worker were vital within the movement.

Physiological well-being in the factory

Initiatives to improve the health and welfare of people through workplace provisions stretched back into the nineteenth century. However, an examination of the annual reports of the Factory Inspectorate in the years prior to the outbreak of the First World War indicate that such provisions were viewed as interesting but unusual innovations, limited to a handful of employers with a reputation for philanthropy. Thus when in 1911 the Factory Inspectorate detailed welfare provisions at the Cadbury factory, listing its landscaped setting, recreational and educational provisions, washing facilities and medical personnel, they were commenting very much on the exception rather than the rule.[44]

Moreover, industrial welfare initiatives were rarely mentioned outside the reports of the lady factory inspectors, who tended to discuss initiatives designed to improve women workers' moral welfare rather than health care provisions. In 1912, for example, the women inspectors noted six new instances of social welfare work by employers of girl labour, and Miss Escreet described how 'at two mills recently I found charming dining-rooms. The furniture of one included books, basket chairs, a piano, and a sewing machine; the other was gaily decorated with paper flowers.'[45] Another employer sought to counter concerns that the factory environment was unhealthy by providing an open-air rooftop workroom for girl workers suffering from ill-health.[46] It was more common, however, for factory inspectors to bemoan insalubrious conditions than to commend good provisions. The unsanitary conditions of factory toilets were a particularly frequent target of complaint in the pre-war reports of factory inspectors. Noting that women workers frequently resorted to using the toilets for a break in the absence of mess rooms and officially designated rest breaks, Inspector Miss Tracey commented that 'to sit on a floor for a few minutes' talk, as I have often seen, is a rest which under even such horrid circumstances is better than nothing'. Tracey's observations indicate that workers were able to subvert provisions to meet their needs, but she felt the answer lay in the provision of proper conveniences and 'the supervision of a nice woman'.[47]

For many commentators, the main priority of any scheme of industrial health and welfare was to create a clean and hygienic workplace by ensuring that lighting, ventilation and sanitation was of an adequate standard. The Factory Inspectorate, particularly the women's branch, had been pioneering this approach before the outbreak of war: thus the Principal Lady Inspector of Factories stated to the 1903 Interdepartmental Committee on Physical Deterioration that 'progress in health in factory and welfare life is mainly a question of raising the ordinary general hygiene'.[48]

In his 1925 volume, *The Health of the Workers*, Sir Thomas Oliver devoted a chapter to the topic 'What factories are and might be'. Oliver, a professor of medicine at the University of Durham, had devoted attention in his earlier publications to the subject of occupational disease, lead poisoning and the dangerous trades. His decision to publish on the health of workers (in a series entitled 'modern health books') reflected the general trend towards health and away from illness.[49] Indeed, Oliver's decision to subsequently publish on the health of the school child illustrates how this re-examination of the factory from a health angle in the interwar years was interconnected within a broader health promotion movement. Oliver asserted that factories first and foremost should be kept clean, and dust removed regularly as 'daylight and sunlight are essential to health'. The emphasis Oliver placed on lighting, temperature and ventilation recalled the physiological approach adopted by the HMWC. 'Work can only be effectively carried on', Oliver wrote, 'if the surrounding temperature lends itself to the comfort of the operatives.'[50]

Edward Hope's 1923 volume, *Industrial Hygiene and Medicine*, emphasised the importance of hygiene, containing a chapter entitled 'General Hygienic Considerations'. Hope's decision to publish on the subject of industrial hygiene also illustrates how factories had been drawn into a broader movement of health promotion; as Medical Officer of Health for Liverpool and Professor in Public Health at the University of Liverpool his background was in the field of public health. 'The influence of housing upon health has for long been recognised', he wrote, 'but the effects of working conditions have received tardier acknowledgment.'[51] Hope advocated clean, well-lit, effectively ventilated factories with good washing facilities and a supply of drinking water.

Endeavours to improve the hygiene of the factory environment were however hampered by the wording of legislation which left legal standards open to interpretation. Until the passing of the 1937 Factories Act, factory conditions were regulated by the 1901 Factory Act which stipulated that ventilation should be 'sufficient', temperature should be 'reasonable', factories should be kept in a 'cleanly' state, 'adequate means' should be found to drain floors and sanitary accommodation should be 'sufficient and suitable'. This made it difficult for inspectors to secure convictions, as Factory Inspector Rose Squire recalled: 'it has often been found difficult to convince a bench of justices of the peace that in prosecuting an obdurate employer one was not merely "fussy" or thirsting for an orgy of spring cleaning at his expense'.[52] These concerns about hygiene mirrored contemporary anxieties regarding cleanliness in the home, and housewives were urged to keep homes free of germs by utilising new technological appliances and choosing easy-to-clean floors, surfaces and walls in kitchens and bathrooms.[53] In 1925, the growing use of vacuum cleaners within factories for walls, ceilings and shelving was noted.[54]

Throughout the early 1920s, the Factory Inspectorate argued that the hygienic condition of factories required much improvement, noting the neglect of cleaning in many premises and the diverse standards of toilet accommodation. The trade problems in the early 1920s hindered the improvement of workplace conditions and restricted the introduction of welfare provisions as employers were unwilling to meet the costs of labour and materials in an uncertain market. Lighting also remained an issue. An international illumination conference held in 1931 argued that no real improvement had been made since the 1913 Departmental Committee on Factory Lighting.[55] In 1936 the Factory Inspectorate expressed grave doubts that the small employer would improve standards unless compelled to do so.[56]

Legal requirements stipulated a minimum standard of medical and first aid provision in interwar factories. A handful of large companies went further than this, inaugurating fairly comprehensive schemes of health care for their employees. These medical facilities, such as the surgery provided by the Queen's Engineering Works in Bedford which is depicted in Figure 1, were frequently lauded within the pages of the annual reports of the Factory Inspectorate and *Industrial Welfare*, the journal of the Industrial Welfare Society. Rarely, however, was the extent of such facilities actually quantified.[57] This perpetuated the misleading impression that factories were amply supplied with medical amenities.

The HMWC had recommended a range of provisions, including first aid cupboards or local dressing stations where minor injuries could be treated by trained workers with sterile dressing and a central dressing station for the treatment of more serious injuries, which should include a surgery, a nurse's room, a rest room and a store room. Such provisions were largely instituted to counter the risk of injuries sustained in the workplace becoming infected, including injuries to the eye.

Within large factories, arrangements could become more complex. The HMWC described how one national filing factory which employed 7,000 staff had constructed an ambulance building, a ward with eight beds for workers, an emergency operating room, a cupboard for surgical appliances, two consulting rooms for doctors, a waiting room for patients and a store room. This factory employed two doctors, one sister and four nurses, and owned its own ambulance wagon to take workers to the local hospital. Workers suffering from ill-health could avail themselves of a nearby convalescent home, run by the Young Women's Christian Association and subscribed to by the workers, while the factory employed a health visitor for home visits to sick workers. Other firms employed dentists to provide free treatment to their workers in the belief that sepsis and infection of the teeth and jaw could allow poisons to enter the body. However, the quality and breadth of first aid provision during the First World War varied between factories. While some factories were equipped with model ambulance rooms, factory inspectors viewed both equipment and personnel in most as inadequate: 'the equipment which I saw, where there was any, was disgraceful. It generally consisted of a bandage, a bit of lint or wool – usually very dirty – and some useless sticking plaster'.[58]

The maintenance of first aid boxes remained a complaint for the Factory Inspectorate in the immediate post-war years; Legge noted disapprovingly in 1923 that 43 per cent of boxes failed to meet required standards of dressings.[59] Under the 1924 Workmen's Compensation Act, factories were required to install a first aid box of a prescribed quality.

A PIONEER FIRM.

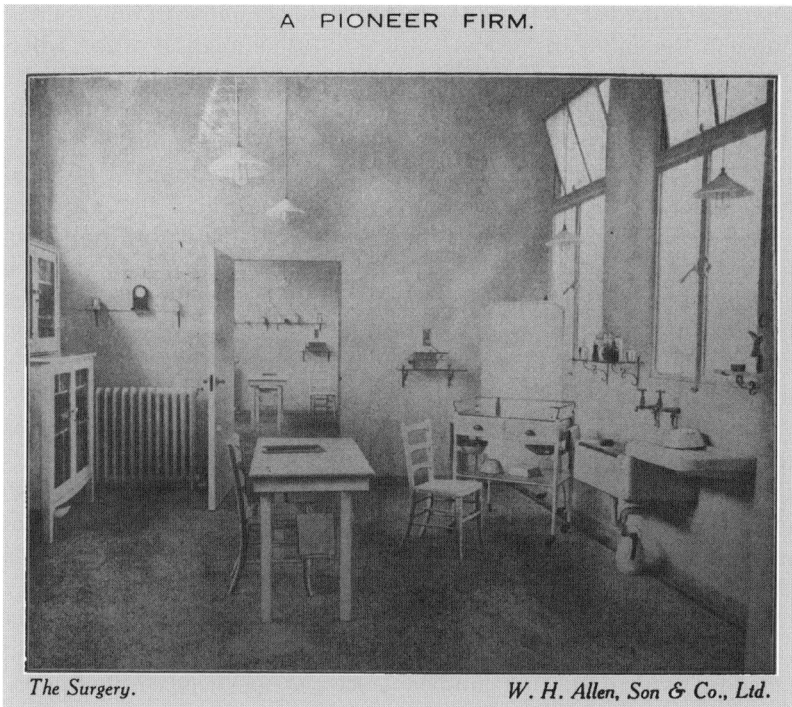

The Surgery.   *W. H. Allen, Son & Co., Ltd.*

Figure 1 'A Pioneer Firm: The Queen's Engineering Works Bedford', *Boys' and Industrial Welfare Journal*, 2 (1919) 5–9. Photo of Factory Surgery, 7

Legge reported that these provisions had largely met with a good response from employers, though the First Aid Inspector Mrs Bridge did cite the case of one firm who sought an exemption from the provisions of the Act on the grounds that the men they employed 'were too dirty and too ignorant to apply a sterilised dressing properly'.[60] In 1924, works doctors were still reported to be a rarity and health provisions were generally overseen by welfare supervisors, trained nurses and St John Ambulance brigade trained men. The requirements of the 1924 Workmen's Compensation Act appeared to fuel the interest of employers in health provisions and the number and standard of ambulance rooms was noted to have increased. In 1925, Mrs Bridge visited 450 factories which had applied for or received a certificate of exemption from the provision of first aid boxes, which could be granted if an ambulance room was provided. Of these 450 factories, 109 employed a fully trained nurse in the ambulance room, with a full-time medical officer working in a further 16 firms.[61] Compared against figures collected by the Factory Inspectorate in 1919, in which 290 ambulance rooms were reported, of which 85 were staffed by a fully trained nurse,[62] there is clear evidence of growth in the number of medical facilities and trained personnel.

The growing range of health provisions available within some factories from the mid-1920s were described in the annual reports of the Factory Inspectorate, industrial health textbooks and in the pages of *Industrial Welfare*. The provision of dental treatment schemes for workers and ophthalmic surgeons began to increase from the mid-1920s while the use of violet ray lamps was first mentioned in 1927. In 1937, Factory Inspector Evelyn Sanderson listed medical facilities such as the provision of artificial sunlight, X-ray installations to detect fractures, fracture and rehabilitations clinics, and the appointment of doctors, dentists, opticians, chiropodists and masseurs. Washing facilities stretched beyond the provision of wash basins to encompass plunge, shower and foot baths, which could help prevent dermatitis.[63]

We need to approach these impressionistic accounts with some caution and contextualise the growth of ambulance rooms and medical personnel within the total number of factories operating within Britain. An indication that voluntary initiatives were transforming only a tiny handful of factories into healthy spaces can be gleaned from a memorandum prepared for the General Council of the Trades Union Congress (hereafter TUC) in 1930 by Sir Thomas Legge, which outlined some of the difficulties of securing the health of workpeople. Legge stressed the limitations of current legislation, grumbling that the National Health Insurance Act was something of a misnomer and that 'National Disease Insurance Act' would be a more fitting title for the 1911 National Health Insurance Act, 'as cure and not prevention is its aim'.[64] This was a damning assessment in the interwar years when health promotion not curative medicine was the ideal. The memorandum argued that there was an insufficient number of medical personnel to safeguard worker health, noting that only around 50 of the 150,000 factories employed a full-time medical officer, while the five medical factory inspectors were incapable of policing conditions within these factories and the 130,000 workshops. Moreover, the 1,700 certifying factory surgeons who examined all young workers between the ages of 14 and 17 within the first seven days of their employment were constrained by the terms of the 1901 Factory Act to listing any disabilities which might preclude a young person from working full-time. Legge wanted to see the work of certifying surgeons taken over by the National Health Insurance Act, with some panel doctors undertaking medical supervision in factories and an emphasis on prevention rather than cure.

It was only in 1937 that the provisions governing conditions within factories were substantially revised.[65] The 1937 Factory Act differed to previous legislation, which had sought to achieve the flexibility required to cover the whole of industry by framing requirements in general terms and adapting them to meet the circumstances of each case. The 1937 Act sought to make requirements more precise and detailed, with latitude stemming from powers to grant exemptions and impose additional restrictions where it was felt to be necessary, placing more responsibility in the hands of the Secretary of State and his advisers. John Bridge, the Chief Medical Inspector of Factories, argued that the Act advanced the all important role of preventive medicine in industry, pointing to powers granted to the Secretary of State to make arrangements for medical supervision where he had reason to believe that cases of illness were due to work. Young workers could now be examined not only to ascertain that they were not unfit to work, but to help determine the work for which they were suited, a process which would be facilitated by the availability of school medical records for this purpose.

Bridge, who had been appointed Senior Medical Inspector in 1927, remained sceptical that improving workplace conditions would lead to an improvement in the health of workers. He frequently used his sections of the annual report to muse on the relationship between health and occupation, arguing that ill-health could be wrongly attributed to working conditions, that conditions in factories were healthier than in workers' own homes, that there was insufficient data on the health of the general as opposed to insured population to build up a picture of the health of the nation and that more knowledge of the relationship between health and work would be beneficial. 'Undoubtedly industry does produce ill-health in certain instances,' Bridge admitted, 'but with some experience of industrial conditions I would say that it is outside of the factory where most of the ill-health arises.'[66] Nevertheless, he believed that medical supervision within the factory could help combat sickness among employees, regardless of whether that sickness derived from the workplace or the home. Bridge favoured a uniform system to record and compare sickness rates in industry. He argued that work could benefit health, pointing to the health problems associated with unemployment.

Another indication of the growing interest in the healthy factory was the establishment of industrial museums by European governments to promote safe and healthy working environments. In 1912, the factory inspectors had helped the Finnish Industrial Safety Museum secure exhibits from British employers, and expressed the hope that 'we may before long watch in our country the beneficial results of general access to similar information in a British Museum of Industrial Safety'.[67] Two years previously, the Treasury had sanctioned the establishment of an industrial museum to illustrate methods of securing the safety, health and welfare of industrial workers. Completed in 1914, the building was used during the war for other purposes. Possession of the building was obtained in 1926 and the Museum was opened the following year; it was subsequently enlarged in 1936.[68] The ground floor contained demonstrations of machine fencing, the gallery level included exhibits on industrial diseases, welfare and fire prevention while the basement demonstrated methods of ventilation, lighting and heating, and contained examples of washing and sanitary accommodation and clothes storage. It is significant that one exhibit singled out for attention by the Factory Inspectorate was a piece of shaft and clothing of a man who had received fatal injuries. This was noted to have attracted a lot of interest among visitors, demonstrating the greater immediacy and visible impact of industrial accidents over industrial disease and health promotion.[69] The Home Office Industrial Museum was part of an international movement, and other industrial safety museums operated in Germany, Italy, Hungary, Sweden, Denmark and the Netherlands. This international collaboration was a feature of the interwar years: in 1925, Dr Bridge took part in the first interchange of visits by medical inspectors of factories from different countries arranged by the International Labour Office, visiting Geneva and Brussels.

The Factory Inspectorate established the Museum as part of its move away from the use of punitive methods such as taking employers to court in order to impose legal minimum standards. Instead the Inspectorate encouraged industry to improve its own standards. With no new substantial legislation dictating minimum health and welfare standards within factories until the 1937 Factory Act, the Factory Inspectorate may have seen voluntary methods such as the Industrial Museum as the most effective means to achieve this.[70] The 1932 Report pointed to the Museum as 'a practical indication that the Factory Department is more ready to help in the betterment of industrial conditions than to seek penalties for breaches of the Act'.[71] Intended primarily for employers, workers and their organisations, the Inspectorate hoped that the

Museum would also prove beneficial for the designers of factories and the makers of machinery and protective appliances, for welfare and social workers and for the students of engineering schools of universities. Inspectors described how employers arrived with contravention notices to find out how they could best comply with the instructions from district inspectors, and the inquiries received from firms looking to improve safety after accidents. It appears, however, that the Museum was not as successful as it had been hoped in attracting workers and trade unionists. The Factory Inspectorate wrote a series of letters to the TUC seeking its assistance in publicising the Museum among trade unionists, but although the TUC featured the Museum in its *Industrial Review* and *Industrial News*,[72] the Inspectorate recorded that visits by these groups had not matched expectations.[73] The Factory Department offered to give lectures to trade union conferences and trade councils about the Industrial Museum. Not all lecturers were well received: in 1933 the Deputy Chief Inspector of Factories wrote to the TUC's Social Insurance Secretary J. L. Smyth complaining that in one instance a factory inspector had arrived to give a lecture, to find that only three officials had turned up.[74] By 1934, a rise in visitor numbers was noted from the average of 5,500 of previous years to 9,500, an increase attributed in part simply to leaving the front door open during opening hours which had encouraged more workers to visit,[75] and nearly 90,000 visited the Museum in its first ten years.[76]

Psychological well-being in the factory

At the same time that factory inspectors sought to persuade recalcitrant employers to meet minimum legal standards, some factories introduced new features designed to enhance workers' well-being. A number began to experiment with colour schemes: one company described in the 1920 report 'have made attempts to carry out effective colour schemes by colouring walls or painting iron work…in cheerful colour or pleasant shades of green',[77] while in 1925 one inspector noted 'the advantages to be derived from pleasant surroundings', commenting on the tasteful colour schemes, plants and ferns in a worsted mill.[78] By 1930, inspectors explained similar initiatives in the new language of industrial psychology, commenting in the *Annual Report* that lime washing had been replaced in many instances with colour schemes of green and primrose, 'in line with prevalent psychological theories', and noting the appearance of flowers and plants around the external surroundings of mills in the North of England.[79] Paint firm Jenson & Nicholson capitalised on the psychological interest in colour in the 1940s, issuing a promotional book entitled *The Function of Colour in Factories Schools and Hospitals*. This work advised companies that the oppressive feeling of heavy machinery could be countered by painting the walls and ceiling a shade of 'receding pale green' and suggested that a cheerful but not overly bright or stimulating colour scheme be adopted for the factory canteen, as employees were likely to arrive at the canteen 'in various degrees of physical and mental fatigue'.[80]

The 1920s and 1930s witnessed a boom in the provision of welfare facilities within factories, and a proliferation of publications by the emergent industrial welfare movement. *Industrial Welfare*, the official organ of the Industrial Welfare Society, published countless glossy spreads on new model factories in its ever-expanding pages, replete with health, welfare and leisure facilities.[81] In the early 1920s, attempts to adapt the factory environment were often undertaken in a rather ad hoc way. Dr Jaeger's Woollen Company, for example, established a makeshift canteen in a corridor which linked one department of the works to another, and portioned off a rest room from the welfare supervisor's office, which was 'painted with a pretty cream distemper, and the windows are hung in the same dainty style as in the canteen. Deep wicker chairs invite the worker to rest for this half hour, and on the tables are magazines, books and games.'[82] Similarly, photographs of ambulance and rest rooms in the early 1920s suggest that they were simply curtained off from production space; the caption to the photograph of the ambulance room at the engineering firm Mather and Platt noted that 'the large workshop can be seen through the window'.[83]

As new factories were constructed in the 1920s, the practice of establishing canteens and other welfare provisions in under-utilised spaces within the factory gave way to the creation of purpose-built spaces or entire blocks, adjacent to the production spaces of the factory, which incorporated a range of welfare facilities. In 1926, for example, *Industrial Welfare* profiled the new club building constructed by the carpet manufacturer James Templeton & Co, which contained a recreation hall, a kitchen and an adjoining dining room for women and girls. Such purpose-built provisions were most commonly found among firms with a large female workforce: two-thirds of the 3,000 workers employed by Templeton & Co were women.[84]

'Welfare now means something more than a third-rate canteen', wrote Factory Inspector Evelyn Sanderson in 1937 as she summarised the wide scope of provisions which could be encompassed by the term 'welfare'.[85] Sanderson argued that even small employers were investing in welfare provisions, which she attributed to the general levelling up of the standard of living, the education work of the Industrial Fatigue Research Board and of safety and welfare organisations and 'the clear evidence that Welfare "pays" '.[86] Sanderson noted that more comprehensive schemes were developing on the new trading estates. Even in years where unemployment was much higher and competition for workers less of an issue, the reports of the Factory Inspectorate stressed that welfare amenities were of increasing importance if employers wished to appeal to the best workers, enabling firms to attract from a wide area. Under the terms of the 1937 Factory Act, many matters which had previously fallen within the remit of discretionary welfare provisions, such as washing facilities, drinking water, accommodation for clothing and seating facilities now became compulsory in all factories.

It was not only the environment in which work took place but the nature of the work which was undergoing change. The shift away from skilled labour towards production processes accelerated in the interwar era, provoking concern among industrial health and welfare experts about the psychological impact on workers of repetitive, unskilled work.[87] John Bridge, the Senior Medical Inspector, expressed concern about the growing mechanisation of processes and the decline in craftsmanship. He believed that 'vastly more days are lost from vague ill-defined, but no doubt very real, disability due to *ennui* than from all the recognised industrial diseases together'.[88] Industrial welfare advocates stated that welfare schemes could help alleviate these problems by making the factory a more beautiful and stimulating environment. Colour schemes and artwork had been advocated in the 1920s; by 1937 gramophones and wirelesses were reported to be in use in many factories for short spells, especially where work was unskilled, light or monotonous: 'It tends to create a cheerful atmosphere', one factory inspector noted, 'and thereby to benefit the health and consequently the efficiency of the workers.'[89] Eldred Hallas, from the National Union of General Workers, argued in *Industrial Welfare* that factories 'should be as beautiful and spacious as possible. Too many of them are places that stunt the body, warp the mind, and destroy the soul.'[90]

An alternative was to provide sources of satisfaction and interest in workers' lives outside employment, and demonstrate to workers how their job contributed to the bigger picture, in an effort to counter the evils of specialisation. Opportunities for outdoors recreation to counteract the detrimental impact of monotonous toil were dependent on working hours. For much of the interwar period, the 48-hour week was standard practice in many trades. Five-day working weeks became more common, and were advocated on health grounds; greater benefits were ascribed to long weekends of leisure which facilitated recuperation from work. This was seen as particularly valuable amid fears of 'speeding up' in working practices. The Factory Inspectorate also stressed the potential for self-improvement, arguing that as standards of education among workers had risen and their interests had widened thanks to lectures given at institutes, free Saturdays were now of greater value.[91]

As industry began to revive in 1935, Inspector Emily Slocock expressed concerns that the longest hours of work were undertaken by the weakest section of the workforce as the healthier workers took employment with companies which offered better working conditions. She also noted that long working hours had become a source of complaint among workers: 'many of the complaints received about long hours are found on investigation to relate to the hours worked by men. Very long hours are often worked and complainants are sometimes surprised to hear that there are no legal restrictions.'[92] Under the 1937 Factory Act, legal hours of work for adult women workers and young persons fell to 48 hours (from the previous limits of 60 hours a week in non-textile factories and 55 and half hours in textile factories), while working hours of young workers under the age of 16 were reduced to 44 hours a week. Workers' opportunities for leisure expanded further with the passing on the 1938 Holidays with Pay Act.[93]

Some employers established sporting and recreational provisions to enliven workers' free time. Proponents of industrial welfare viewed recreation facilities as a means to alleviate boredom and monotony and maintain workers' health, morale and output. Schemes varied with regard to such elements as who ran and controlled the facilities, the voluntary or compulsory aspect of participation and the cost to the worker or employer of participation. Facilities listed included company bands, picture houses for remote factories, team sports, swimming pools, rambling and cycling clubs, gardening, indoor games, dancing, gymnastics, pianos, company dramatics societies, libraries and literary societies. A few firms even organised holiday camps and tours abroad. Objectives which by now appear rather familiar underpinned these schemes; they attracted and helped retain workers; they advertised the firm; and potentially they could maintain and enhance workers' physical health and state of mind, promoting efficiency and employee contentment and reducing the risk of industrial unrest.

Publicising new ideals of industrial welfare: advertising the model factory

As new forms and practices of industrial welfare began to be espoused, many commentators sought to outline their vision of the ideal factory. These accounts reflected the diversity of opinion which existed regarding the purpose of industrial health and welfare developments and were written to serve a variety of purposes: to publicise the work of employers engaged in industrial welfare; to promote government policy, to attract potential employees and to advertise a particular company to consumers. These latter two objectives were intertwined; consumption was made possible by those manufacturing goods, while the same workers formed part of the mass market for consumer goods. This point has been elaborated upon in some detail by Miriam Glucksman, who argued that women were caught up in a circuit of production and consumption which linked the workplace and the home and transformed domestic labour.[94] As housewives began to purchase items which they previously would have made, advertising featured the image of the homely factory to convey the home-made, healthy character of the products. While accepting that many companies hoped to attract employees with the promise of good working conditions, it seems inconceivable that this factor alone could explain the new emphasis placed upon the health and welfare of employees and the new model factories within which they worked. It would appear that this information made the product more desirable in the eyes of the consumer, either because of consumer demand for ethically produced goods, reflecting the fact that many consumers were themselves industrial labourers with a natural interest in working conditions, or because goods produced within such a working environment were believed to be superior, produced with more care in modern, hygienic surroundings by clean, healthy employees.

Both these sentiments lay at the heart of the co-operative movement in this era, which offered a means for working-class women to challenge capitalist business interests and reject substandard goods through their consumption choices. By 1920, 1,379 co-operative distributive stores were in operation in Britain, supported by more than 4.5 million members: annual sales of the English Co-operative Wholesale Society reached £105 million in the same year.[95] The success of the Co-operative Wholesale Society in this era may well have influenced other companies to advertise their products in a similar fashion, promoting the quality of their products and the working conditions under which they were produced.[96] These advertising strategies ensured that the healthy factory had become marketable by the interwar years, a desirable object with promotional potential.

This led some observers to question the purpose of what they viewed as extravagant and unnecessary aspects of some modern factories. In her account of welfare work, Dorothea Proud argued that the 'straining after splendour' evident in the outward facade of many factories was ill-fitted to function, suggesting 'amusement or luxury rather than monotonous toil'. Many factory gardens, she claimed, 'have an obtrusiveness which suggests that they concern the public more nearly than the employees'. The growing practice of 'using the process of manufacture for advertisement', Proud observed, had induced a number of companies to install visitors' observation galleries into their factories.[97] At the Lever Brothers' soap factory in Port Sunlight, for example, gangways and galleries enabled visitors to observe employees at work in the different departments.[98] Similarly, pharmaceutical firm Boots integrated visitor-friendly features into its new factory in Beeston in 1933, creating public pathways through spaces of production. A handbook for guides produced by the firm described a route for visits through the factory over gangways, bridges and balconies which enabled visitors to watch various production processes.[99] Proud believed that innovations such as company bands, sports teams and works magazines were vulnerable to exploitation as 'advertising media' by cynical managers. Even the most public spirited employer, she wrote, must be tempted to 'soften the crudities which are bound to result from the efforts of untutored workers'.[100]

Proud herself was engaged in a promotional activity, namely the advertisement of welfare work, and her critique of what she described as the excesses of the modern factory was incorporated to bolster her argument that industrial welfare supervision was an expert profession which no modern efficient businessman would be without or should attempt on his own. Nevertheless, employers expressed concern with the welfare of their employees needs to be critically interrogated. Proud described how a factory might be 'approached by marble steps between growing plants', and yet concealed within one could find underground workrooms.[101] John Walton takes this argument to an international level, arguing that paternalist employers were 'a conspicuous and well-documented minority' and emphasising the incongruity between William Lever's model village for employees, Port Sunlight, and Lever's utilisation of indentured labour on his palm oil plantations.[102] Even Rowntrees, a Quaker-founded firm with a strong anti-slavery heritage, exploited its colonial producers: welfare facilities for the firm's cocoa producers were largely a chimera, with profits channelled back into British developments and facilities.[103]

Many advertisements of model factories in the 1920s represented the industrial workplace as a homely arcadian idyll which offered its workers all the comforts of home.[104] Such representations were often reinforced by descriptions of the pastoral company village built to house the firm's workers. Company housing schemes emerged against a backdrop of labour unrest and became a site in which the expertise of social reformers, welfare supervisors and architects came into conflict.[105] Many companies loosely modelled their company towns on the ideals of the garden city, seeking to resolve the tension between the pastoral idyll and industry explored at the beginning of this chapter. Developed and popularised by Ebenezer Howard in his 1898 volume *To-morrow: A Peaceful Path to Real Reform*, which was revised and reissued in 1902 as *Garden Cities of To-morrow*, the garden city movement sought to simultaneously resolve the problems posed by rural depopulation and urban sprawl. Howard proposed small, planned towns in the countryside with public gardens and parks integral to the plan, encircled by a rural estate.[106] Factories would be on the perimeter of the towns in designated zones. Howard oversaw the construction of two garden cities in Britain. The first, Letchworth, was inaugurated in 1903. Welwyn Garden City was initiated in 1919 in the era of post-war reconstruction. Befitting a garden city, Welwyn was designed with its own public services, shopping amenities, good housing and places of employment, enabling people to work and live within easy reach of the countryside and have their own garden and open space.

The ideal of the garden city embodied significant health implications. The inhabitants of the overcrowded, unhygienic industrial centres, it was argued, were less healthy than their rural counterparts and therefore less effective workers. A study undertaken by the Industrial Fatigue Research Board in 1927 on the physique of women in industry reflected these assumptions, stating that the evidence collated for the study 'supports the general belief that the rural population is of better physique than the urban'.[107] By scientifically planning new towns which allowed inhabitants to enjoy the benefits of the countryside while still offering easy access to industrial employment, these problems would be alleviated. While the garden city movement, therefore, had roots in late nineteenth-century utopian socialism, it also stressed the health-giving properties of nature, fresh air and sunshine: ideas that were at the core of the interwar health movement. It was at the same time progressive and regressive. Moreover, healthy employees and healthy surroundings featured as marketing points in the advertisement of industrial products in the interwar years. In its implementation, the environmental aspects of Howard's ideals were stressed rather than the social reform latent in his proposals, just as William Morris's practical suggestions for factory improvement were widely adopted while his political objectives were discarded. Stephen Ward has argued that the longevity of the garden city movement can to a large degree be attributed to its diverse characteristics, enabling proponents to apply aspects of the movement selectively while discarding what they found unappealing.[108]

Indeed, the success of the garden city cause owed much to the willingness of leading industrialists George Cadbury and William Lever to associate their factory villages, Bournville and Port Sunlight, and garden factories with the movement. In their turn, Cadbury and Lever used the rhetoric of the garden city to market their companies. Houses in Port Sunlight sported an eclectic variety of facades inspired by different epochs in British history, embodying Lord Lever's aim to construct 'an *ersatz* version of Merrie England'.[109] Port Sunlight, noted historian Brian Lewis, made 'good advertising copy... It smiled out of Lever Brothers' posters, a radiant, gleaming, sun-blessed idyll.'[110] Chocolate manufacturing firm Cadbury had been a pioneer in this field, relocating from its production site in central Birmingham to a new location several miles from Birmingham in 1879, which it christened Bournville.[111] Cadbury consciously represented its model factory at Bournville as a reinvention of the pastoral idyll,[112] persistently using the image of the 'factory in a garden' in advertising throughout the late nineteenth and early twentieth centuries. One typical advertisement from the 1890s emphasised the purity of Cadbury cocoa by stressing the hygienic manufacturing process. The advertisement informed readers that the firm's cocoa was 'made in that most pleasant of manufactories – the factory in a garden – of Cadbury Brothers, at Bournville' and featured illustrations of girls in white dresses seated at benches packing chocolates, the waterfall and bridge at Bournville and a view of the factory with trees in the foreground. Other advertisements explicitly connected the processes of production and consumption. One pamphlet took the reader on a journey from production in the garden factory to consumption within the home, folding out to reveal sequentially a tin of cocoa and a family seated at their dining table, accompanied by text which read 'health and happiness comes to the people who drink Bournville cocoa made in the factory built by Cadbury in the garden village of Bournville'. Another pamphlet featured a young girl drinking a mug of cocoa, accompanied by the text: 'Cadbury's cocoa is pure; do not give your children foreign preparations mixed with alkali.' The pamphlet opened to reveal the factory girls of Bournville dressed in white frocks and playing games outdoors, and sitting in a semicircle which encompasses trees and the factory windows, blending the factory interior into its garden exterior.[113] Both pamphlets emphasised the healthy, local nature of the production process, concealing the international dimensions of chocolate production.[114]

From 1910 onwards, Cadbury issued visitors with a pamphlet, 'Bournville: the Factory in a Garden', which advertised the factory in a garden and life in the Bournville factory village. 'Situated in one of the healthiest and prettiest parts of Worcestershire', an early edition explained, 'everything has been done to beautify the factory and its surroundings, the buildings themselves have been planned with the object of realising the most modern hygienic ideals'.[115] The cover shown in Figure 2 comes from a later edition which advertised both the factory in a garden, including its medical provisions, gardeners, catering facilities and 14,000 square feet of windows which were cleaned daily, and life in Bournville village itself.

Figure 2 'Bournville: The Factory in a Garden', undated pamphlet. Reproduced courtesy of Cadbury UK

Cadbury even sought to elevate its production processes to the realm of children's fiction in a series of pamphlets, derivative of *Alice in Wonderland*, which recounted the adventures of Elsie and the Bournville bunny. 'I am here to guide the little boys and girls who want to go through the beautiful workrooms at Bournville and see how Cadbury's cocoa and chocolate are made', the bunny explains to Elsie in 'Bunny: Visit to Bournville'. 'You will see that everything is kept very, very clean in these rooms so that nothing nasty or bad can possibly get into the cocoa or chocolates.' Further on in her tour, Elsie discovers that 'it is not "all work and no play" at Bournville' as the bunny shows her the firm's playing fields, girl's swimming bath and recreation ground. Cadbury clearly deemed this so successful that it merited two sequels: 'A Visit to Sunny Cocoa Land' which paints a picture of happy cocoa producers in the tropical paradise of Trinidad, and 'Bournville Bunny', in which Elsie witnesses the girls' lunch break and visits the main office.[116]

While Cadbury enjoyed a reputation as a pioneer in health and welfare provisions for its workers, a number of other companies also exploited the ideals of the garden city to promote their products. An article written in 1918 to advertise the engineering firm Vickers drew attention to the company's welfare provisions and bucolic setting on Walney Island. It discussed the houses, gardens and hostels built by the company and the welfare provisions established by the firm, which included a club, reading and games rooms, a pub and a theatre. The most striking feature of the piece however was the juxtaposition of industrial production alongside images of rural life and recreation; the island was eulogised as a pastoral retreat amid the ravages of war, replete with allotments, outdoors sporting facilities such as a golf course, the beach, a bowling green and tennis courts:

> Down on the beach children built sand castles, men and girls were bathing, and scores of people sat reading on the pebbles or watching the swimmers. And a mile away, almost hidden by the swell of the island, the chimneys of the shipyard and munition shops belched forth the smoke that told of unceasing industrial activity of vital importance in the prosecution of the War.[117]

A 1914 advertisement for Crosse and Blackwell conjured the image of a homely factory to promote the company and its products, seeking to romanticise and domesticate both the factory settings and the relationship between employees and employers. It described the 'old-fashioned amity' between the employers and their 'happy and healthy' workers. Within 'the extensive kitchens...a small army of white-coated young women' were described working,[118] while a walk through the jam and preserve factory 'reveals no intricate machinery or novelty in manufacture. Here we see jam and marmalade made in the good old-fashioned home-made way ... this part of the factory may well be taken as a model for the average kitchen.'[119] This advertisement combined old-fashioned paternalism with modern technology; the description of the employees contained militaristic and scientific imagery, yet the production process was described as old-fashioned and home-made: the factory was at once domesticated and paternalistic yet also clean, modern and scientific, mirroring developments in 1920s hospital architecture in which historical styles created a conservative exterior while interior design revealed a 'technological fetishism'.[120]

The advert sought to draw together two further incompatible threads: the 'efficient working of a huge concern', it claimed, 'is only made possible by perfectly harmonious relations between employers and employees'. The growing scale of manufacturing was typically seen in this period to have torn asunder the personal bonds between employee and manager; here it was claimed that the bonds could be restored. However, the secret ingredient ascribed to securing good relations within this firm, paternalism, would soon lose favour within industrial practice. As industrial firms grew in size and complexity, industrial philanthropy, which had characterised nineteenth-century industrial welfare practice, was displaced by a new model of industrial efficiency which systematically integrated welfare into management practices.[121] The 'moral obligation' expressed for the girl workers would soon lose favour, though the particular needs of women in industry continued to receive attention by welfare supervisors.[122] In 1914, the figure of the girl worker was considered to attract enough sympathy for the advert to regale its readers with a description of the numerous provisions inaugurated on her behalf: a girls' dining club, a savings bank, a swimming club, a dramatic club, an orchestra and athletics club.

An advertisement placed by the hosiery firm I. R. Morley in the same month also publicised its history of philanthropic welfarism. The general tenor of the piece however evidenced new trends which would predominate in the interwar years. The reader's attention was drawn to the 'airy, sun-bathed work-rooms' and the 'scientific organisation of plans for the betterment of those employed'. The advertisement claimed that welfare provisions, which included dining, reading and smoking rooms, sporting, recreational and education facilities and a hostel for young workers, had not been established out of any sense of moral obligation, but the recognition of 'modern factory managers' that such facilities were a profitable investment. It is evident from this advertisement that the happiness and quality of the firm's employees were regarded as a significant selling point. Not only did welfare facilities 'improve the grade of labour', by making the work more attractive, 'it also increases the output and improves the quality of the work itself'. Within the advertisement, the company interwove the quality of the product, the workers, output levels and company ethos into one marketable package.[123]

It is notable that many firms which chose to promote their products in this way manufactured food items. Advertisements placed in the 1920s for Ovaltine, Chivers' Jams and Lyons, for example, all promoted the image of idyllic rural factories, well-cared-for workers and home-made products.[124] Nutrition was an important topic in the interwar years: on the one hand, increasing emphasis was placed upon the benefit of a balanced diet if health was to be maintained while at the same time anxiety was expressed that people on a low income, let alone those on benefits, were unable to afford a healthy diet.[125] These unpalatable realities were emphasised in 1933 when the British Medical Association's Nutrition Committee published a report on dietary requirements which set out the minimum income required to purchase a healthy diet.[126]

Not all garden factories were inspired by visions of the past. A 1926 advertisement for Ovaltine, headed by an illustration of the new factory the company had commenced building in 1924, illustrates how the ideal of the homely factory began to be displaced by new modernist visions. Depicted against a backdrop of rolling country hills and sited within formal gardens, the large windows and art deco facade of the factory appear unabashedly modern and somewhat at odds with the bucolic surroundings and the horse and cart portrayed in the foreground. Captioned 'The Home of Good Health', the advertisement described the beautiful healthy surrounding countryside before focusing on the factory: 'the factory itself is the ideal of what a factory should be. Spotlessly clean, full of sunshine and sweet country air, and surrounded by gardens to make a happy and healthy staff – such is the home of "Ovaltine".'[127]

Similarly, the new Shredded Wheat factory at Welwyn Garden City also blended a rural setting and factory gardens with a modernist structure. An advertisement placed for Shredded Wheat in 1937, depicted in Figure 3, encapsulates how new model factories functioned to entwine consumption with production in the interwar years. Here, the modernist Shredded Wheat factory is at the very heart of the advertisement and has been brought into the home of the consumer, printed on the box containing the product and placed next to a bowl of the product in a domestic setting. The linkage and interrelationship between product, factory, consumer, employee, nation and health are clearly made. The box featured in the advert carries the slogan 'Britons make it – it makes Britain' while the main slogan of the advert, 'the food that brings the crowning blessing of health and fitness to all' implicitly interrelates

Figure 3 Shredded Wheat Advertisement, *Cavalcade*, June 1937 reproduced courtesy of Société des Produits Nestlé S.A.

the health of the employee with the health of the consumer, bringing the factory into the home. The crown symbolically links the consumer through the product to the monarchy, and by extension, national identity: as Thomas Richards asserts in his study of advertising in the Victorian era, 'the best way to sell people commodities was to sell them the ideology of England'.[128]

The Boots Beeston factory offers one of the most striking examples of the influence of modernist architecture on factory design in the inter-war years. Boots, like Cadbury, had a history of philanthropic provision that stretched back into the nineteenth century. Between 1920 and 1933 the firm was under the ownership of the United Drug Company of Boston, and in 1927 a Works Planning Committee was established to consider current fashions in industrial architecture. Opened in July 1933, the new Boots Beeston factory was described in press reports as an 'industrial crystal palace' and 'factory of utopia'. The purpose of this factory was to promote the firm and the consumption of Boots products as much as it was to provide a site for the production of goods. Thomas Richards has argued that the Crystal Palace, explicitly alluded to in this description and still standing until its destruction by fire in 1936, elevated commodities by creating a new spectacle of exhibition which would shape Victorian advertising. Rather than isolating production from consumption, Richards asserted that the Crystal Palace 'integrated the paraphernalia of production into the immediate phenomenal space of consumption'.[129] Constructed using concrete, the Beeston factory exemplified the modernist interwar factory, providing its workers with ample sources of daylight and sun through its large windows.[130] Owen Williams, the architect who designed the factory, believed that functional concerns should be uppermost, describing factory buildings as 'the shell surrounding a process'. 'I can picture the factory of the future as a great single span shell housing a vast machine with its workers dotted about in no way that can be related to definite horizontal planes or floors', he told an audience in 1927.[131] Williams later deployed the same flat-slab construction method to design one of the flagship buildings of the interwar health movement: the Pioneer Health Centre in Peckham. Both buildings were constructed to maximise sunlight and minimise the need for internal walls, materialising the synergy of ideas between industrial rationalisation, health and modernity.

Boots used its new factory and model working conditions to advertise the brand in two souvenir pamphlets published in 1934 and 1936. The 1934 pamphlet stated that Owen Williams had 'become sensitive to the living, productive organism which is "Boots" and designed a home for it of such a nature as to develop, rather than repress, the united personality of this body of workers'.[132] The desire to reflect the personality of the employees as a whole may nevertheless have effaced individuality. Engineer David Yeomans and architect David Cottam regarded the four-storey-high central atrium as at once an 'awe-inspiring' and 'dehumanising' space, 'whose scale and efficiency reduce the individual worker to insignificance'.[133] In its conception, the new Boots factory was unabashedly modern; rather than attempting to integrate the factory into a rural setting, Boots chose to emphasis the disjuncture between the idyllic pastoral scenes of the past and the new modern factory it had constructed. In this vein, one Boots publication described the setting of the factory from the imagined perspective of eighteenth-century local poet Henry Kirke White. It described how Clifton Grove 'rises above the River Trent, on a well-wooded cliff side, with a noble avenue of trees sweeping up the crest. At the apex...stand the old Hall'. Today, 'looking down on those verdant meadows, we shall see the new factory of Boots'. What, the author asked, would Kirke have made of this change?

> Would he have been stirred to wrath against desecrating man, or, in keeping with the spirit of modern poetry, would he have been moved to wonder? For here is no factory in the sense of that word, once a synonym for an industrial prison; here rises a shining edifice whose lines carry the eye from plane to plane in perfect harmony.[134]

In 1936, the TUC's Medical Adviser Dr Hyacinth Morgan filed a report on a visit he had taken to the Boots factory at Beeston where he had been given a guided tour by the Chief Medical Officer for Boots, Dr Lockhart, who was also a member of the Medical Research Council and the Industrial Health Research Board. In his report, Morgan approvingly noted the medical facilities available to workers in the factory. Dr Lockhart was assisted in his work by two nurses and alongside treatment, consulting and rest rooms, he had facilities to provide sun ray and violet ray treatments. Morgan reported that workers were well supervised and precautionary measures were taken in the manufacture of hazardous substances. He noted that the ventilation and heating arrangements within the premises were good, and that the structure of the building made it possible to supervise workers on any floor, without commenting on whether this extended disciplinary control and surveillance over the workers. Boots, he decided, exercised 'great paternalism over the welfare of the workers' and was 'allied to the workers' interests as far as can be yielded under present conditions'.[135]

Worker demand for healthy factories

When the factory inspectors noted improvements in the sanitary conditions of many factories during the First World War, they attributed it to competition for workers, suggesting that the environment of the workplace had become an important factor for employees alongside wages and hours of work. The establishment 'of a modern factory, with bright and cheerful rooms', they reported in 1914, 'has often the effect of raising the standard in the locality'.[136] By 1933, one factory inspector reported that welfare facilities, especially mess rooms, had become so widespread throughout industry that complaints were often received regarding their absence in industries where they were not legally required.[137] In 1937, the Factory Inspectorate received 3,098 complaints, covering 5,329 subjects. Of these subjects, 866 were deemed to be outside the jurisdiction of the Factory Inspectorate, but most of the remaining 4,463 related to hours of employment, ventilation, sanitation and temperature, of which about 50 per cent were substantiated on enquiry.[138]

The assertions of factory inspectors that welfare facilities were a prerequisite for employers who wished to attract high-calibre employees need to be rigorously evaluated. Nick Hayes, in his analysis of workers' reception to welfare provisions on British building sites between 1918 and 1970, argued that welfare amenities such as canteens and mess rooms threatened to undermine the masculine, independent, tough identity that site workers had constructed for themselves. As a result, the provision of welfare amenities was greeted apathetically by workers on the sites, even though the unions campaigned for such provisions.[139] Certainly, the reports of the Factory Inspectorate in the immediate postwar years detailed instances of resistance by male workers to welfare provisions. Advocating works committees in 1921, Hilda Martindale complained that the response of workers to welfare orders had been disappointing, particularly in the case of men.[140] A year later, Miss Ahrons discussed the 'disappointing' response of returning male workers to welfare orders which had been instituted in oil cake mills to assist the welfare of women workers who had entered the trade during the war. Ahrons complained that while the structural requirements of the order had been complied with, they had been subjected to rough treatment by the male workers: 'Everywhere … she has heard the same tale of destruction of the arrangement provided for the workers' comfort'.[141]

It would be misleading also to equate these welfare developments with improvements in workers' health. As we have seen, the inauguration of medical facilities beyond the minimum legal requirement of a first aid box affected only a very small proportion of factories. Even the partisan Industrial Welfare Society acknowledged that sometimes the 'frills' of welfare had been accomplished while basic health problems had been left untouched. Within *Industrial Welfare*, welfare supervisors increasingly cited the goals of employer-driven welfare schemes as taylorism, efficiency and human management; in other words, welfare had become a new means to increasing managerial control over workers. Joseph Melling and Helen Jones have both argued that the development of welfare facilities by employers was motivated primarily by a desire to secure workplace discipline.[142] Melling's analysis emphasised how employers utilised industrial welfare in the period 1880 to 1920 to undermine trade unionism by providing alternative financial schemes for savings, targeting groups of workers who played a crucial role in the division of labour.

In her study of interwar welfare schemes, Jones argued that employers were able to delay new legislated standards of health and safety within factories by promoting voluntary welfare schemes which in practice led to a wide discrepancy between model conditions and workplaces which barely conformed to the requirements of the 1901 Factory Act. A report issued by the TUC on health provisions provided through employers' welfare schemes in 1932 acknowledged that the earliest health provisions introduced were beneficial and that much remained to be done in smaller workplaces where working conditions were very bad. However, the TUC argued that the development of increasingly elaborate schemes in individual firms was 'bound to retard progress towards that unified system of complete and adequate national health service which must be regarded as the ideal'.[143] Indeed, as the example of the Industrial Health Education Society in the next chapter demonstrates, while employer organisations may have paid lip service to industrial welfare, most seemed keen to obscure real health issues and evade further legislated standards and responsibilities.

Conclusion

Because the history of the healthy factory is not simply one of development but decline also, it can be surprising to find that in the 1920s and 1930s factories could exemplify modernism and health and be looked upon as a key site of health production and an arena in which to base a nationalised health scheme. This new vision of the factory transformed the interior of the factory to incorporate spaces and facilities associated with the home such as gardens, bathrooms, kitchens and rest rooms; it challenged beliefs about the functions of a factory and the role of management.

It is evident when examining representations of the factory in the interwar years just how embedded it was in the interwar health movement; one only has to look at the synergy of ideas regarding the health-giving properties of fresh air, light, aesthetics, exercise and nutrition. This representation of the factory suggested that the health of industrial workers was primarily a matter of general health; it offered employers a discourse through which to proclaim their interest in workers' health while downplaying the risks posed by industrial disease and accidents and suggesting that responsibility for ill-health lay with the workers themselves. Even William Morris, whose extensive critique of the dehumanising nature of the factory system of production featured earlier in this chapter, used arsenic in the production of his wallpapers and was indifferent as to the health consequences.[144] It should also be stressed that the promotion of the healthy factory was not primarily driven by medical professionals; while the Association of Industrial Medical Officers was formed in 1935, the actual number of medically qualified staff working within factories, aside from the certifying factory surgeons, was tiny.[145] Industrial medicine barely featured on the interwar medical curriculum,[146] while the reports of the medical factory inspectors still focused on the prevalence of industrial disease. Rather, the image of the healthy factory was promoted by the profession of industrial welfare supervisors and advertisers who had found a new way to market products.

The defining feature of most of the trends of the interwar years examined within this chapter – the health movement, the garden city, industrial welfare and industrial development – was the conflicting impulses inherent at the heart of each. The garden city was both regressive in its pastoral arcadianism, but progressive in its attempts to humanise industrial and urban life; likewise the health movement looked backwards to an idyllic past when health was enhanced by man's union with nature, while advocating ambitious plans to modernise health provisions. Older industries were declining while new industries were emerging. For some firms, the Depression meant a tightening of the budget and a cutback of welfare expenditure; other firms expanded their range of provisions hoping to entice the best workers. Attempts were made to humanise the factory environment at the same time that mass production methods eroded skill and satisfaction levels of factory work. Industrial firms and welfare supervisors advertised provisions within the best factories while resisting efforts to improve standards nationally through legislation; evidence indicates that these ideal conditions existed in a minority of factories only. To fully explore the limitations and constraints faced by those seeking to make the factory a healthy place we need to analyse the fiercely contested political debate about the provision, purpose, control and funding of health services in the interwar years: this is the subject of the next chapter.

Another discernible trend in the interwar years was the gradual ascendancy of psychological approaches towards workers' welfare, which slowly eclipsed the physiological model adopted by the HMWC. This shift can be seen in the declining number of features in *Industrial Welfare* which lauded ideal welfare facilities and the increasing number devoted to methods of scientific management, a transformation reflected in the reconfiguration of welfare supervision into personnel management. Employers and personnel managers began to argue that efficiency was compromised not because workers' physiological comfort had been neglected but because due attention had not been paid to whether they were temperamentally suited to the work or whether it met their intellectual and emotional needs.[147] This change should not be overemphasised; nevertheless, attention shifted from the provision of physical amenities within the factory to the psychological impact of repetitive and monotonous work on health. By the 1930s the model of the sleek modernist factory eclipsed arcadian visions of the homely garden factory. The potential offered by industrial psychology to alleviate mental health problems which arose when workers were placed in work unsuited to their capabilities or experienced stress and trauma following workplace accidents will received further consideration in Chapter 4.[148] What is relevant to note at this point is that this new approach to the workplace curbed further innovations. After all, why bother to install a library or bowling green at a factory if the key to industrial efficiency lay in placing the worker into the correct job, especially if workers were inclined to question the motives which lay behind said bowling green? This is not to dismiss the developments that had occurred out of hand; many provisions which had formerly fallen under the rubric of voluntary welfare facilities were now enshrined in law through the 1937 Factory Act. Moreover, the complaints received by the Factory Inspectorate testified to worker demand for canteens, mess rooms and good standards of hygiene.

The healthy factory was initially promoted by the industrial welfare movement, which claimed that the establishment of welfare facilities would humanise industry and re-establish the personal touch between employer and employee which had been lost as the size of industry escalated. As welfare supervisors retreated from close supervision of the factory's welfare and sanitary provisions, they left the door open for the nascent professions of industrial medicine and personnel management. Two trends would subsequently emerge. On the one hand, concern with the welfare of particular groups of workers persisted and Chapter 4 will explore how efforts were made to tailor industrial work to the personal needs of young workers, women and employees with particular health needs. On the other hand, the new medicalised approach promoted by industrial medical officers envisaged medical supervision of the individual worker and the workplace environment to promote healthy working conditions and prevent illness from occurring. The success of this model remained fundamentally linked to the numbers of personnel available and the future shape of a national health service.

3
Taking Responsibility: The Politics of Industrial Health

The politics of health in the interwar years

A brief exchange of letters between Dr Hyacinth Morgan, Medical Adviser to the Trades Union Congress (TUC) and Dr Howard Collier, head of the Department of Industrial Hygiene and Medicine at Birmingham University between 1935 and 1939, encapsulated several themes which shaped and constrained political interactions in the field of industrial health: how the intended functions of trade unions, medical organisations, charitable groups and employer associations determined the extent and nature of their involvement; how the unequal financial resources of different organisations shaped activity; and finally, the disappointment and misunderstanding that could ensue when local organisations felt that they had not received the support they deserved from national bodies. In a letter sent in 1944 to Morgan, Collier asserted that his Department would not have been forced to close had the TUC provided financial and moral support. By failing to provide a subscription to the Department, in spite of appeals for assistance from local labour leaders, Collier claimed that the TUC had lost an 'incomparable opportunity', and he criticised 'the inadequate efforts of the TUC to help working people in the matter of industrial illnesses'.

Collier argued that all trade unions and every large branch should appoint part-time industrial medical officers. 'The large factories do this', he noted; 'why not the Trade Unions?'[1] Morgan informed Collier that he had failed to grasp the functions of trade unions and the TUC. 'It is true that local labour leaders frequently put up requests to the TUC', Morgan wrote, 'but after all, the TUC is a national body dealing with a matter of nearly five to six million workers and they cannot be expected to accede always to the demands of local...labour leaders.'

Moreover, Morgan claimed, the activities of trade unions were constrained by finances and most branches could not afford to employ doctors. Even if they could, Morgan continued, the resulting medical service would be one-sided and would not advance the development of industrial medicine. 'You say "the large factories do this" and then you say "why not the trade unions"', Morgan wrote. 'The trade unions have different functions to factories.'[2]

The previous chapter explored the growing interest in health promotion as opposed to simply treating illness. Initiatives to refashion the factory into a site of health production were one aspect of this broader interwar movement of health improvement. Attention was increasingly paid to creating a healthy factory environment, in the belief that if workers ate nutritious food at staff canteens, were provided with ample washing facilities, were given a clean, light and well-ventilated workspace and offered sport and recreational opportunities, they would in turn participate in maintaining their health. An economic rationale underpinned these interventions; preventing illness was cheaper than treating it.

Different health care models proliferated in the 1920s and 1930s and debate raged as to where healthcare provisions should be based, what services should be provided, who should be entitled to these health services and, crucially, who was responsible for funding, organising and running such services. This chapter explores the flourishing interactions between different groups in the interwar years as they instigated, debated and contested different schemes to improve the health of factory workers. As indicated in the Collier–Morgan correspondence, the contributions made by different organisations to the promotion of industrial health need to be situated within the broader context of the sphere of political activity in interwar Britain if they are to be effectively assessed. However, to examine these developments solely from the perspective of health is problematic. 'Health' was, and remains, an ambiguous term with no fixed definition which could be employed to describe often contradictory objectives. In practice, many of the initiatives which were identified with health dealt as much with illness.

The exploration of wartime industrial health initiatives, undertaken in Chapter 1, outlined factors which shaped the responses of the State to the development of industrial health. Jurgen Habermas has argued that the collapse of the public sphere in the twentieth century left the State to arbitrate between the competing interests of different social groups.[3] Habermas traced the emergence of a social welfare state, which circumvented the role of public institutions to intervene more directly than ever before in the lives of its citizens. Keith Middlemas's interpretation of twentieth-century political interactions in many respects parallels Habermas's argument. Middlemas asserted that the State circumvented the threat of political upheaval and class strife in the early years of the twentieth century by incorporating employer associations and trade unions into governance in the years 1916 to 1926. The cooperation and collaboration of these organisations created a new political harmony which calmed class conflict and would last into the 1960s. Middlemas believed that the compromises made by these organisations to ensure the success of this strategy had undermined democracy, weakening the legitimate field of parliamentary operations and stifling dissent.[4]

More recent commentators have urged us to reconsider the role of organisations, localised groups and institutions, suggesting that the influx of private interest groups into governance may have enhanced democratic participation. Nancy Fraser's reformulation of the public sphere provides a helpful way to examine the role played by interest groups, which she termed '*subaltern counterpublics*', parallel discursive arenas that enabled members to formulate counter identities and discourses.[5] She suggested that the participation of interest groups, far from circumventing the democracy of the public sphere, helped to contribute to it by enabling a broader cross-section of people to participate in debate.

While the multiplicity of political agents active in the formulation of policy and practice in the twentieth century offered the potential for greater democratic participation, the anxieties expressed by Middlemas and Habermas are not without foundation. Fraser, for example, noted that 'counter-publics' were not always democratic, often operating their own methods of exclusion. One question which this chapter will address, therefore, is the extent to which the TUC managed to represent the interests of workers divided geographically and occupationally. Was it responsive to the needs of local branches, or did it, as Middlemas asserted, suppress discontent within the ranks as it sought to establish itself as an estate of the realm?

As this chapter will demonstrate, the impetus for the development of industrial health policy did not always lie with central bodies: many innovative approaches and solutions were instigated by peripheral trade union branches in tandem with local practitioners. Contributors to a recent volume edited by Steve Sturdy emphasised the role played by institutions around which multiple medical publics developed.[6] Sturdy argued that the continual formation of local publics, which participate collectively to encourage government activity through both debate and forms of collective action, discourages the notion that the public have been reduced to passive consumers. 'We impoverish our understanding of the public and its functions', Sturdy warned, 'if we confine ourselves to thinking solely in terms of discourse, and thereby neglect the role of institutionalised action in the constitution of the public sphere.'[7]

Middlemas interpreted the new moderate and professional approach adopted by the TUC after the First World War as undemocratic. But it is also important to acknowledge that the contested politics of industrial health did not take place on a level playing field. Interest groups exerted different levels of influence and authority on the subject of industrial health. Following Bourdieu, one would expect that industrial workers lacked the necessary class background (habitus) and skills (capital) to persuade employers, medical practitioners and government officials to adopt their proposals.[8] The TUC's strategy could be interpreted in this light as an attempt to augment the capital and habitus of the labour movement and thus enable the TUC to compete more successfully and participate on a more equal footing.

The TUC was, after all, struggling to represent the interests of its members in what Harold Perkins described as a society increasingly structured around professional status and career hierarchy.[9] Perkins argued that professions emerged which offered trained, expert services to society, seeking in return status, income and authority. As different professional groups competed for public resources, charitable organisations petitioned the State for funding for deserving causes and trade unions sought to protect their conditions of employment, the State became the mediator for disputes of interest.[10] If the twentieth century was indeed the century of the professional expert, to what extent could a non-professional organisation, such as the TUC, participate, unless it refashioned itself? At this point, it is pertinent to draw on Joseph Melling's suggestion that rather than thinking solely in terms of the interests pursued by respective organisations, or of how responsibility and blame should be apportioned between organisations in the field of industrial health, we should consider what constraints they faced.[11]

It was within this broad political framework that the objective of improving the health of the workers was pursued and contested by a variety of organisations with overlapping memberships. For all the groups involved, the promotion or obstruction of industrial health was only one objective amid a much wider agenda which varied by organisation. For the TUC, the promotion of industrial health formed part of a broad social agenda which it pursued alongside its traditional role of coordinating trade union action on questions of wages, hours and working conditions; here I would concur with Melling that there was no inherent conflict of interest between the pursuit of these agendas.[12] To enact their agendas, organisations could focus on achieving short- or long-term objectives; they could coordinate their campaign with other organisations if they were prepared to compromise on goals, or they could choose to prioritise one objective over another, either because they assigned greater importance to it or for pragmatic reasons.

The economic situation of the interwar years certainly called for a pragmatic approach. Rising unemployment in the interwar years undermined the numerical strength of both the TUC and the British Employers' Confederation (hereafter BEC), compromising their claim to represent the interests of workers and employers.[13] Arguably, concern about the impact of work and the working environment on health was eclipsed by concern about the effect of unemployment and poverty on health, demonstrated in the psychological studies undertaken on the effects of unemployment and the British Medical Association's (hereafter BMA) 1933 study on the minimum expenditure on food necessary to maintain a healthy working capacity which raised questions about levels of unemployment assistance.[14] Wartime industrial production had provided the impetus to government intervention in industrial health. When the country slipped into depression and levels of unemployment rose, the political will to improve health drained away in light of more pressing problems, particularly as the budget available to the government and employers became restricted. State and privately run schemes could be downsized and short-term objectives had to be rethought if they were to be achieved. One response of the State to a decline in the economic situation was to push responsibility for the protection of workers' health into the private sector, urging trade unions and employers to work together to provide solutions.

Key players

The main organisations which contested the future of industrial health in the interwar years were employer associations, trade unions, medical organisations and voluntary groups. This chapter focuses on the perspective of the TUC and will investigate how adequately it represented the interests of its members and how effectively it could compete in the political field of health care in an era when expert professional knowledge, especially in the field of medicine, was privileged. Furthermore, the chapter will explore the extent to which the TUC considered the promotion of health and the development of health care services for workers and the population at large to be one of its legitimate functions, and what initiatives and approaches it considered to be beyond its remit.

Trade union membership stood at 2.5 million in 1910, of which around three-fifths were affiliated to the TUC. Membership increased dramatically through the war, reaching 4 million by 1914, 6.5 million by 1918 and a record 8.5 million in 1920. Trade union strength in this period was demonstrated not just by increasing membership but through militancy: strikes were widespread and trade unions coordinated action to maximise impact. Militant action ceased for a period after the outbreak of war but resurfaced in 1915, prompting the government to establish the National Labour Advisory Council and to institute a system of compulsory arbitration through the Munitions of War Act. It was no coincidence that the Health of Munition Workers Committee (hereafter HMWC), discussed in the first chapter, was established in 1915 at the height of labour militancy. In 1916, the new Prime Minister and erstwhile Minister of Munitions, David Lloyd George, appointed a number of Labour Party and trade union representatives to government posts.[15] Undeniably, trade union action made the government focus on workers and working conditions, yet the TUC itself did not play a central role in these developments. TUC historian D. F. MacDonald argued that the war 'enormously enhanced power and public prestige of the unions', but notes that that it simultaneously 'gave free play to the dissident elements within the movement'.[16] The larger trade unions commanded a significant amount of power, while much of the strike action during the war was unofficial, orchestrated by shop stewards and at odds with the more moderate central trade union leadership.[17]

Between 1920 and 1938 the TUC sought to become a powerful political force which would be capable of representing the opinions of the labour movement to employers and the State.[18] This process was facilitated by changes in personnel and organisation, which began to occur in the early years of the 1920s. From 1869 until 1921, the main body carrying out the work of the TUC was its Parliamentary Committee, which met infrequently and was responsible for lobbying sympathetic politicians, organising deputations to government departments and reporting back to Congress. In 1921, the TUC transferred the executive functions of the Parliamentary Committee to the newly established General Council, entrusted to resolve disputes between unions and promote common action by the trade union movement on general questions such as hours of labour and wages. The position of General

Secretary was made a full-time post for the first time in 1922 with the appointment of Fred Bramley.

Following Bramley's death in 1926 Walter Citrine was appointed as General Secretary to the TUC, continuing in office until 1946. Under Citrine, the TUC engaged in a policy of damage limitation, casting off its radical approach and attempting to entice workers from all political parties to join. This was an era marked by professionalisation of practices and efforts to engage in dialogue with employers. Following the collapse of the General Strike in 1926, in which it had assumed a central role without securing consent from all unions, the TUC was unwilling to overstep its boundaries and sought to develop policy slowly through dialogue, negotiation and consent.[19] Subsequently, the TUC stressed its desire to seek moderate solutions through constitutional means. In 1928, the General Council initiated a series of talks between the TUC and industrialists to discuss subjects such as factory legislation, works councils and security of employment. The talks were eventually to flounder in the face of employer indifference, but they point to the TUC's interest in this era in forging a collaborative approach to negotiating improvements for workers.[20] This approach was also evident in the TUC's health policy. Other shifts in TUC policy reflected the evolution of the organisation. Industry-wide collective bargaining began to replace localised settlements. As part of Citrine's efforts to create a moderate and responsible image for the trade union movement, the TUC began to target middle-class workers, incorporating technical, administrative and supervisory staff. Similarly, the interwar years saw the TUC concentrate on providing services for those with less wealth, rather than campaigning to redistribute wealth.

The expansion of full-time personnel at head office, the acquisition of modern office equipment and Citrine's decision to initiate a central registry for TUC files in the 1920s enhanced the efficiency of office procedures.[21] Milne-Bailey's 1934 book, *Trade Unions and the State*, reflected recent developments within the TUC. Milne-Bailey argued that the TUC was becoming a semi-public institution with interests and duties extending beyond employment issues.[22] TUC health policy largely postdates the collapse of the General Strike and was therefore characterised by caution and moderation, though this moderate approach was not one shared by all affiliated trade unions or by local branches. The appointment of former Senior Medical Factory Inspector Sir Thomas Legge as the first Medical Adviser in 1930 and the burgeoning interest in health issues thus reflected both the growing tendency to import expertise from outside the trade union movement to develop the credentials and standing of the TUC, and the increasing range of issues which the TUC felt rightly came within its remit. The development of a coherent TUC policy on health issues was also facilitated by the Secretary of the Social Insurance Committee, J. L. Smyth, who remained in office throughout the interwar years. By the mid-1930s, growing membership began to add credence to the TUC's claims to represent the voice of workers; between 1937 and 1938, net affiliated membership grew by half a million.[23]

Notwithstanding these developments, it is important not to overestimate the strength of the TUC and to acknowledge the limitations on the power of this organisation. While the TUC may have been examining a range of subjects hitherto considered outside its sphere, its potential for action remained constrained. Affiliated unions were keen to preserve their autonomy and resisted any moves to cede power to the TUC, which remained incapable of controlling the actions of individual unions. Furthermore, the trade union movement was weakened by 'poaching' – competition between trade unions for members. The interests and concerns of local union branches could differ from the official stance of their national union, let alone the position of the TUC. The combination of economic depression and the collapse of the General Strike, which had been orchestrated by the TUC, proved detrimental to membership levels, which declined from 8.3 million in 1920 to 4 million by 1933.[24] This drop was not simply a result of rising unemployment: the percentage of unionised workers within the workforce also declined. While 45.2 per cent of the workforce were unionised in 1920, by 1930 union density among the male workforce was 30.8 per cent, and among women workers 13.4 per cent; the total of 6.3 million trade union members in 1939 represented a density of 38.9 per cent and 16 per cent of the male and female workforce respectively.[25] The 1927 Trade Disputes Act and Trade Unions Act, which outlawed solidarity actions or strikes that were designed to coerce the government by enforcing hardship on the community, remained on the statute book throughout the interwar years. Despite its ambitious expansion of activity, Robert Taylor has argued that the TUC remained more a pressure group than an estate of the realm in this era.[26] Membership only started to recover by 1939 when it reached 6 million, returning to the 8 million mark by 1943. In his study of the TUC, Taylor assessed that it was the role played by the organisation during the Second World War which 'transformed the TUC into more of an Estate of the Realm than at any other time in its history'. The National Joint Advisory Council established in October 1939 contained an equal number of representatives from the Ministry, the TUC and the BEC. Ernest Bevin, General

Secretary of the Transport and General Workers Union, was appointed Minister of Labour and National Service by Churchill in May 1940. He subsequently established a Joint Consultative Council comprising of seven TUC representatives and seven employer representatives which had weekly meetings with Bevin.[27]

During the interwar years employers sought to ensure that industrial health remained a matter for individual employers and voluntary agreements and thus they attempted to block further legislation of working conditions and campaigned against any extension of workmen's compensation.[28] The medical profession also intervened in debates about health services for workers; doctors wanted any future service to reflect their views of how medical practitioners should be employed and remunerated. A report on the future provision of medical services, issued by the Ministry of Health's Consultative Council on Medical and Allied Services in 1920, suggested a national network of health centres and sought to encourage the participation of private medical practitioners in a state service.[29] Recognising the opposition of most doctors to a state medical service, Lord Dawson, the architect of the report, envisaged that such a service could only be established gradually, 'by stealth'.[30]

The structure, functions and approaches taken by the BMA had been under transition in the early years of the twentieth century, as it sought to establish itself as an influential organisation, capable of translating the interests of its members into public policy and establishing itself as an organisation qualified to speak on a range of public issues.[31] Rather like the TUC, the BMA had established a Parliamentary Bills Committee back in 1872 and had relied on a strategy of lobbying Parliament on behalf of the profession, approaching MPs and ministers believed to be sympathetic. Again, like the TUC, the BMA found this approach was no longer effective. This was made very apparent in 1912, when despite strenuous lobbying the government forced through a National Health Insurance Bill which failed to meet the demands of the profession. While Peter Bartrip has argued that the BMA did win important concessions from the government, he also noted that the BMA felt itself defeated, with membership figures plummeting: like the BEC and the TUC, therefore, the BMA's claim to represent the interests of doctors had been undermined. Charles Hill, who was appointed Assistant Medical Secretary to the BMA in 1932 and would become the TUC's main point of contact in the interwar years, argued that the fight over National Health Insurance 'almost destroyed us [the BMA] as a coherent and representative body'.[32] By 1918, the BMA represented only 15,000 of the 43,000 on the medical register, and faced competition from other medical organisations.[33]

As issues of national health moved up the political agenda in the interwar years, the BMA adopted new methods to ensure that it adequately represented the interests of the medical profession. The BMA's efforts to secure better representation of medical interests by establishing a fund to support doctors seeking election to Parliament did not meet with unqualified success. As Roger Cooter has noted, the 72 doctors who were successfully elected to Parliament between 1918 and 1945 were divided along party political lines and membership of different medical organisations, exacerbating divisions within the medical profession.[34]

By the 1920s and 1930s, the BMA, like the TUC, became more involved in questions of public health and welfare. In 1933, the BMA produced a report on nutrition which listed the calorific needs of men, women, children and the elderly and estimated the cost of providing such a diet: this report was produced by a committee chaired by Charles Hill.[35] Two years later, the BMA established a Physical Education Committee to investigate how national standards of physical fitness could be raised by physical training. Around the same time, the TUC was displaying a new interest in subjects such as nutrition and health, as well as matters of direct concern to the BMA such as the nature of the medical curriculum and trade union representation on medical committees and organisations.[36] Whether or not the BMA was simply seeking to dress up professional interests as medical issues and matters of public concern, as Cooter has suggested, its change in policy meant that both the TUC and BMA were campaigning on similar grounds, and that both organisations may have viewed collaboration as a useful means to formulate policy and present a stronger case.

The BMA was by no means the only organisation claiming to represent doctors' interests. Founded at a meeting in 1930, the Socialist Medical Association's principal goals were to campaign for a preventative and curative socialised medical service, the promotion of socialism within the medical services and the advancement of the health of the British people.[37] The Socialist Medical Association (hereafter SMA) actively promoted a vision of a socialised medical service in the 1930s which was strikingly at odds with the objectives of the BMA. Some medical practitioners, mainly those working in the field of general practice, formed the Medical Practitioners' Union, a TUC-affiliated trade union:[38] members of all three of these groups could overlap.

In Chapter 1, the role of the State in the development of wartime industrial medical and welfare services was explored. While the inter-war years were dominated by Conservative rule, the Labour Party nevertheless emerged as a credible governing party which was to hold office twice, in 1924 and 1929–1931; the opportunity for the TUC to enter into dialogue with the State and promote its own perspective on issues of policy thus appeared to be on the horizon.[39] Indeed, four joint policy departments were established with the Labour Party in 1924, but the TUC complained that it was inadequately consulted. Relations between the Labour Party and the TUC during the second Labour minority government were soured by a disagreement between the two organisations about how to resolve economic crisis, the TUC rejecting proposals to cut levels of unemployment benefit. The failure of the Labour Party to ratify the Washington Convention, which would have instituted a 48-hour week and 8-hour day, or to drop the 1927 Trade Disputes Act were also sources of grievance. Closer collaboration may have been thwarted in part by the TUC's decision to adopt an independent political stance, arguing that it worked for all workers, irrespective of their party political stance. Throughout the 1930s, the TUC and the governing Conservative Party remained in intermittent contact and in 1935 both Citrine and Arthur Pugh, secretary of the Iron and Steel Trades Confederation and Chairman of the TUC's General Council during the 1926 General Strike, were knighted.

State policy vis-à-vis workers' health was enacted through permanent officials and medical professionals. Factory inspectors and certifying surgeons, employed to examine all young workers entering industry to ascertain their fitness to work, were first employed to uphold the regulations of the 1833 Factory Act. By 1930, there were 1,700 certifying surgeons as opposed to five medical factory inspectors.[40] Panel doctors were employed under the terms of the National Health Insurance Act to provide basic medical services to workers. The role played by voluntary organisations in this period should also not be overlooked. The Industrial Welfare Society, for example, founded in 1919 by the Reverend Robert Hyde, was nominally a professional organisation run for the emerging profession of welfare supervisors, but with an executive committee comprising almost entirely of the managing directors of major companies. The Industrial Health Education Society (hereafter IHES), which will be discussed later in this chapter, could also be seen as a way for employers to infiltrate into seemingly apolitical and objective organisations. Voluntary organisations played an active role in the interwar politics of industrial health and offered another means through which members of professional and trade associations could intervene.[41]

Finally, it is worth pointing out that people could approach the issue of workers' health from multiple political perspectives. Sir Thomas Legge, for example, began his career as a doctor and became the first medical Factory Inspector. He developed an expertise in industrial health which led him to oppose the government over its failure to ratify the Washington Convention. In his retirement, he took up post as the first Medical Adviser to the TUC in 1930, and subsequently became involved in the work of the IHES as a representative of the TUC. Legge, in the course of his career, was a doctor, a state employee, an industrial health expert, a trade union employee and a member of a voluntary organisation. He was by no means the only individual to occupy such diverse positions.

Medical expertise, trade unions and the operation of health care services

'A medical man', one TUC memoranda noted, 'would have contacts with the Ministry of Health, the Home Office, the Medical Research Council, the BMA, and in a thousand and one ways he would have access to medical publications and experience, both nationally and internationally, which are not open to us as lay men.'[42] This memorandum, written in 1931 when the TUC was searching for a new medical adviser, indicated that the TUC's lack of professional status and expert medical knowledge inhibited its attempts to shape industrial health policy in an era when professional knowledge was seen as something of a prerequisite to those wishing to engage in dialogue with medical organisations. Even the SMA was reluctant to engage in meaningful dialogue with laymen, as John Stewart has noted in his study of the organisation. Somerville Hasting, President of the SMA, asserted in 1933 that 'medical staff must have the sole determining voice as to the type of medical treatment to be given to each patient'. Similarly, in 1938, Hastings stressed that it was 'very difficult for the public to appreciate what is essential for the adequate treatment of disease'.[43]

The TUC attempted to surmount this difficulty in 1930 when it appointed Sir Thomas Legge as the first Medical Adviser to the TUC. This appointment, part of Walter Citrine's strategy in this era to professionalise the work of the TUC, was something of a coup. It helped transform the TUC into a force to be reckoned with by employers and the State. Although the TUC first appointed a medical adviser primarily to assist in the scheduling of diseases for workmen's compensation, Legge and his successor Hyacinth Morgan rapidly transcended this function. In part, the function of the Medical Adviser was to bring to the attention of the TUC medical issues which affected its policy which it would otherwise be unaware of. Another function, however, was to legitimate the concerns already felt by the TUC regarding issues of health and illness which it struggled to convey in a society increasingly stratified by professional expertise. A memorandum produced by the TUC in 1940 described a number of responsibilities undertaken by the medical adviser which included representing TUC interests on committees, maintaining contact with the BMA, keeping abreast of the current literature and institutional provision for industrial medicine, interviewing professional men who visited the TUC on medical matters and advising on any medical or public health query.[44]

Legge had been the first Medical Inspector employed by the Factory Inspectorate in 1898. He was knighted in 1925, and at the time of his resignation from the Factory Inspectorate in 1926, he was a recognised expert in the field of occupational health with many publications to his name, having served on committees and lectured widely on occupational disease in addition to his duties as an inspector.[45] When he took on the role of medical adviser to the TUC in 1930, Legge was aged 67. The robust approach he was to take towards groups whose methods he did not approve of indicates that Legge at this stage felt he had nothing to prove and intended to express his opinion candidly. 'I came into the Factory Department thinking: "oh yes; it would be all right if the workpeople would only wear respirators and wash their hands"', Legge remarked in 1930. He left the Inspectorate in 1926 having arrived at some markedly different conclusions: that poisoning resulted from the inhalation of dust and fumes and that if the dust and fumes were stopped, the poisoning would cease; that efforts to remedy the working environment rather than intervene to protect the worker directly would be most successful and 'that unless and until the employer has done everything – and "everything" means a good deal – the workman can do next to nothing to protect himself'.[46] These sentiments were popularised as Legge's axioms in his posthumous book, *Industrial Maladies*.[47] Thus, despite his short period in office, Legge appears to have transformed the TUC into a body capable of critiquing and rejecting the medical arguments of other organisations such as the IHES, which was at this stage funded by the TUC, and of advancing a TUC line of argument.

A memorandum produced by the TUC after the death of Legge contemplated the functions of the medical adviser and the developing role of the TUC in health issues.[48] This document stressed the need to study the relationship between industries and disease, bemoaning the low status of occupational illnesses within the medical curriculum and academic medicine which resulted and claiming that this led to a relative ignorance among panel doctors regarding the illnesses their patients suffered from. While trade unions sought to bring attention to the diseases which affected their members, they found it difficult to make a scientific case, while every time the Home Office sought to extend the protection of workers it faced powerful opposition from employers. The importance of the Medical Adviser as a mediator between the trade union movement and medical organisations became more apparent throughout the 1930s as the TUC became involved in an array of complex health issues.

Following Legge's death in 1932, the TUC decided to appoint Hyacinth Morgan, who had ten years of trade union work as a medical adviser to the Union of Post Office Workers. Morgan also had strong links to the Labour Party, serving as an MP from 1929 to 1931, and 1940 to 1955.[49] His attempts to forge links between the TUC and medical organisations met with mixed success. Morgan was a member of the SMA, but this did not promote cooperation and collaboration between the latter organisation and the TUC. Relationships with the BMA were also strained. In a letter to Dr Hill, Morgan claimed, 'I have always been anxious to get the great profession of medicine through its recognised and responsible organisations into touch and association with the national trade union movement with its finger on the working class pulse.'[50] However, while Morgan was a member of the Council of the BMA, his position within the BMA was not unproblematic and indeed Hill instigated this exchange of letters to apologise to Morgan for the discourteous reception he had received at the BMA house from Dr Forbes.[51] Morgan brushed this incident aside, writing:

> I wanted to work peacefully and placidly within the profession in my trade union work, all of which was crudely and bluntly bludgeoned as "unethical" according to BMA rules and policy, and not covered by their pronouncements. I have had my desired revenge in showing that there is very good work to be done in more harmonious co-operation.[52]

Ultimately, it proved difficult for Morgan to adopt a stance which would not antagonise any of the groups to which he belonged, though his personal temperament may have also been an influence: Morgan's obituary noted that 'his outspoken criticism of lethargy in official quarters often led him into fierce controversies'.[53] An unpublished letter Morgan dispatched to *The Times* on the subject of sickness in industry demonstrates precisely these outspoken tendencies. Industrial medicine, Morgan asserted,

> ... is in its swaddling clothes in Great Britain mainly because, as one brilliant humanitarian specialist told me recently, 'the palaces of Harley Street are not adorned out of the fees of workmen'. One day the General Medical Council, awaking from its senile drowsiness will realise its duty, and place occupational diseases in the proper place in the medical curriculum.[54]

It was perhaps fortunate for Morgan that this letter was not published as it may well have hindered the TUC's efforts to secure a greater place for industrial health on the medical curriculum. In 1936, the TUC decided to submit evidence to the General Medical Council which was revising the curriculum. The TUC advocated a prescribed special course on industrial medicine and industrial disease and felt that teaching on orthopaedics, industrial accidents and workmen's compensation should also be included. It did not simply wish to see a greater place ascribed to industrial illness within medical education, however; the TUC wanted to see the emphasis shift from disease and curative medicine to a focus on prevention and health. Rather than perpetuating a system whereby doctors were instructed in 'abstruse scientific theories', the TUC believed that trainees should in future study medicine in relation to the social and economic environment. Training 'should be for health and its preservation as well as for the recognition, diagnosis, treatment, alleviation, cure and eradication of disease'.[55] It suggested that the first two years of a medical course should be devoted to the subject of 'normal health' before disease processes were studied. The General Medical Council agreed to TUC recommendations that a special short course of six to eight lectures should be incorporated into the final year of a medical degree as a compulsory subject. Nevertheless, a memorandum produced in 1944 noted that the 42 medical schools in Britain had failed to act upon these recommendations.[56]

Specialist postgraduate training for doctors in industrial health was virtually non-existent in interwar Britain, aside from the work of Dr Collier's ill-fated Department of Industrial Hygiene and Medicine at Birmingham University which had been established from funds provided by local industrial concerns in 1934. The Department offered training to industrial nurses and to qualified medical practitioners who wanted to work as industrial medical officers. A flyer issued for the new Department stressed that 'emphasis will be laid upon the discovery of preventive measures rather than upon the cure of established disease'.[57] When the companies who had funded the enterprise withdrew their funding during the Second World War, the Department was forced to close. Without the support of the TUC, Collier asserted, certain 'very powerful influences' were able to destroy the Department, publicly stating that the Department's work had a 'socialising tendency'. He concluded that industrial disease would never receive the attention it deserved in medical training until a financially independent university department was established.[58]

Morgan dismissed Collier's claims, arguing that what he proposed was beyond the remit of the TUC's functions. However, just three months later, Morgan would contradict himself. Reviewing the unsatisfactory state of affairs in industrial medicine, including the closure of the Birmingham Department and the fact that the only lecturing on industrial health was given by Morgan himself at the Westminster Hospital, Morgan recommended that the TUC should make a gesture to demonstrate its dissatisfaction. He suggested that this could be achieved by either endowing a lectureship in industrial medicine or an industrial disease institute. A permanent endowed lectureship, Morgan felt, would not simply lead to greater recognition among the medical profession of the value of industrial medicine as a specialty; as it would be the first permanent lectureship on the subject, it would 'be a magnificent tribute to the TUC interest in the diseases of industry and the value placed on workers in its proper tuition'.[59] These deliberations were pre-empted when Manchester University established a Chair of Industrial Medicine in 1945. The shortage of doctors with specialist knowledge of industrial health would prove to be a significant stumbling block in the TUC's campaign for a state-run national industrial health service, as will be explored further in Chapter 5. In the context of interwar politics, the general lack of knowledge among the medical profession on the subject of industrial health could only increase the gulf between doctors and trade unionists.

In 1946, Morgan was one of four doctors successfully sued by the BMA for repeating allegations regarding the Association's referendum on the NHS Act, and was forced to apologise and pay legal costs.[60] Morgan's dispute with the BMA was rather unfortunate as the files for the interwar years reveal that while the TUC engaged with a wide range of organisations, it was perhaps most keen to court the approbation and cooperation of the BMA. As the BMA began to develop its work in the field of public health, it found itself working in an area in which the TUC was also increasingly becoming active. Both organisations may have viewed collaboration as a useful means to formulate policy and present a stronger case.

A dispute which erupted in a Welsh village regarding the operation of the local medical service brought the BMA and TUC together and raised questions as to whether laymen were knowledgeable enough to direct and manage medical services, and, in the long term, what role workers could play in the management and operation of the future national health service. A subscription-based medical service had been established in 1910 by trade union groups in Llanelly which was organised by the Llanelly and District Workmen's Medical Committee (hereafter LWMC). The dispute erupted in 1934 when the Committee decided to utilise some of the subscription money currently paid by workers to fund ancillary specialist services, reducing the remuneration received by the practitioners already working for the scheme. The Committee claimed that they had become dissatisfied with the level of service provided and recruited eye, throat and surgical specialists. The existing practitioners, unhappy with the proposed reduced levels of remuneration, established their own rival medical service; in response, the Committee employed three doctors who were prepared to work on a salaried basis to provide medical services for workers and their dependants. The original doctors subsequently brought charges before the General Medical Council against the doctors newly appointed by the Committee, claiming that they had advertised their own scheme and deprecated that run by the former practitioners.

In April 1935, the South Wales Miners Federation wrote to the TUC informing them of the situation.[61] Subsequently, in a meeting with the TUC's Social Insurance Secretary J. L. Smyth and Dr Morgan, representatives of the LWMC argued that if the doctors they employed were struck off by the General Medical Council their service would have to be abandoned as no other doctors would work for the scheme in future. They argued that at the heart of the dispute lay a very important issue as to the rights of workers to run medical schemes, and on these grounds appealed to the TUC for assistance.

Characteristically, Citrine felt impelled to put that matter before the General Council for consideration before any action could be taken. LWMC member J. L. Evans, who wrote expressing fears that the BMA was planning to fight on behalf of the former practitioners and asking for TUC funding to defend the scheme, was directed to circulate a resolution through the unions asking for the intervention of the General Council. Local union branches also approached the general secretaries of their unions asking them to intercede on their behalf with the TUC General Council. Fred Smith, General Secretary of the Amalgamated Engineering Union, wrote to the TUC in May 1935, enclosing press cuttings from the *Llanelly and County Guardian*. One clipping included a summary of a speech given by SMA member Dr Edith Summerskill, who supported the scheme as an example of how a state medical service might operate and bemoaned the conservative nature of the medical profession.[62]

However, the resolutions forwarded by unions to the General Council which they had received from their Llanelly branches were to no avail: a meeting of the Finance and General Purposes Committee on 24 June decided to recommend that no action be taken, on the grounds that the case did not come within the Standing Orders of Congress. 'Everyone appreciated that it [the scheme] is a very good one and highly successful', wrote Smyth to Evans on 28 June, informing him of the TUC's decision. Smyth highlighted the complexities of the case: the refusal of the original doctors to accept the new levels of remuneration could be viewed as a strike, while the Medical Practitioners' Union, which had recently affiliated to the TUC, had members on both sides of the dispute. Smyth sought to offer some comfort to Evans, claiming that both the BMA and the former practitioners were sick of the dispute and wanted the matter settled, while the General Medical Council was far from keen to adjudicate.[63] There may well have been some truth in this statement: the doctors employed by the LWMC to replace the original doctors lodged counter charges in July 1935, alleging that the original doctors had also advertised for patients. A later summary of the affair produced by Dr Morgan noted that as the evidence on both sides was virtually identical, the General Medical Council would either have to acquit all doctors or strike them all off. The case would also have highlighted constitutional problems as all the elected representatives of the Medical Council were BMA nominees, thus bringing into question the ability of the Council to act in a judicial capacity in the case.[64]

Further interviews held by the TUC with the BMA and the LWMC suggested a desire on the part of both parties to achieve settlement. The TUC adopted the role of mediator; Morgan liaised with Dr Hill on behalf of the BMA and Smyth and Tewson negotiated with the LWMC. At an unofficial conference in October 1935, chaired by Smyth, an agreement was reached between the two sides which compromised between the competing claims of medical professionals and trade unionists to control medical schemes which were organised and paid for by workers. It was proposed that a new medical service be established under a management committee composed equally of doctors and representatives of the subscribing lay members. In addition, a medical sub-committee and a lay sub-committee would also be established. Finally, a central committee would handle any disputes. The previous doctors were to be reinstated and the doctors employed by the LWMC after the dispute had broken out were to have their contracts terminated. This scheme appears to have been drawn up by Morgan, and initially placed before the BMA at an unofficial meeting in April 1935 as a possible settlement to the dispute.[65]

The episode illustrated the important role the TUC was seen to occupy as a negotiating body, not only within the labour movement but to professional bodies such as the BMA. The TUC's decision to appoint a medical adviser was no doubt a significant factor in this state of affairs and Morgan assumed a leading role in the Llanelly negotiations.

However, the BMA was not the only medical organisation with which the TUC had to contend when negotiating a settlement of the Llanelly dispute. Dr Welply of the Medical Practitioners' Union argued that the medical side of the central committee should be composed of members of his union and not the BMA. The Medical Practitioners' Union also tried to capitalise on the affair, claiming credit in the national press for settling the dispute. In a letter to Smyth, Morgan bemoaned Dr Welply's attempts to gain credit for his organisation for negotiating the settlement, without paying any tribute to the work done by the TUC officers.[66]

The ongoing wrangling between individual practitioners and the Management Board documented within the TUC files, continuing through to the implementation of the National Health Service Act in 1948, demonstrates the difficulties experienced by the TUC in negotiating a settlement which would prove acceptable to the interests of all.[67] After Morgan was sued by the BMA in 1946, J. L. Evans wrote to Morgan asking if there was a defence fund to which the Llanelly and District Medical Services could contribute to assist Morgan's legal battle against the BMA. Evans no doubt relished the boot being on the other foot, and this intervention demonstrates the deep-seated resentment simmering between the local labour groups and the BMA.[68]

Although not an unmitigated disaster, it could be concluded that the TUC's involvement in the settlement of the Llanelly affair was far from successful. Acting as a mediator proved to be a rather thankless task and TUC personnel were called upon long after the supposed settlement to adjudicate on small disputes. It could also be argued that the TUC had let down individual unions, failing to endorse the visions of worker-controlled health services envisaged by the LWMC and conceding too much ground to the BMA. This is largely the interpretation Ray Earwicker arrived at. He asserted that the relationship between the TUC and BMA in this era was 'characterized by trade union deference on questions of professional competence', that 'the definition of this competence was broadly interpreted by the BMA representatives as a means of securing the BMA point of view' and that, consequently, efforts to achieve consensus were a failure.[69] While there is much to support this argument, it fails to take into account the limitations placed upon the TUC's capacity to act by unions themselves, the difficulties faced in trying to resolve a dispute when trade union affiliated members were on opposing sides, and the complexities of negotiating a settlement which would prove satisfactory to the BMA, the Medical Practitioners' Union and the LWMC. While the TUC may not have secured an outright victory for the LWMC, its intervention saved the practitioners employed from being struck off the medical register and ensured the survival of the scheme. Moreover, the TUC's moderate stance and tactics won support from the BMA, enabling a closer relationship between the two organisations. In 1936 the BMA and TUC established a Joint Committee to resolve disputes between medical practitioners and groups of workers in the wake of the Llanelly dispute.[70] The Committee considered subjects such as the National Health Insurance Bill, the 1937 Factory Bill, the possibilities of a National Maternity Service, the cost of living, rehabilitation of injured workmen, and the possible shape of a future national health service. Both TUC staff and BMA members presented memoranda to the Committee, and some joint memoranda were authored.

The SMA also sought to enlist the support of the TUC for its agenda in the 1930s. In January 1932, Dr Bushnell of the SMA had an interview with Smyth on the subject of workers' health, in which Bushnell criticised the work of the IHES, stating that their lectures were not very useful from a labour point of view, and attacked elements of the medical system, such as the waiting lists for hospitals and the inadequate GP service.[71] Bushnell wanted the TUC's assistance with statistics and research to assist the SMA's promotion of a socialist medical service and health education for workers. Although not outwardly hostile, Smyth's response was rather guarded, pointing out the TUC's own work in the area of health and Smyth's belief that only a socialist government would institute a socialised medical service.

Not rebuffed by this lukewarm reaction, the SMA sent a copy of a report advocated a state medical service to the TUC in 1933. The Report urged a service which was free for all, which was concerned with preventive as well as curative medicine, and which was based around the medical team rather than the individual doctor. It suggested the introduction of scholarships to enable children from secondary schools to achieve a medical qualification and a revision of the medical curriculum to provide more preventative health training. The Report envisaged a network of health centres which would include specialist services and departments for child welfare, light treatment and X-rays, dental clinics and clinics for early mental disease. The Report suggested a single hospital system dominated by general hospitals, noting that 'it is especially important that the Factory Health Service should no longer be separated from the general health administration of the nation'.[72]

The TUC's response towards the SMA throughout the 1930s appears to have been characterised largely by apathy. Although the TUC medical advisers met with SMA members to discuss the Association's plans and initiatives, the TUC did not commit itself to supporting the SMA. After 1938, the TUC's attitude towards the SMA degenerated into outright hostility and it encouraged affiliated unions to avoid the SMA's initiatives. The General Council of the TUC was primarily concerned that trade unions or trades councils attending SMA events would endorse a policy on health which might be at odds with the future policy of the TUC or Labour Party. Local groups, trade council and individual unions appeared to be either unaware of this potential hazard, or indifferent. In 1942, the SMA wrote to the TUC offering to provide specialised knowledge and assistance to individual unions which would target the specific problems of particular industries. Morgan advised the TUC not to become involved. He argued that the SMA hoped to 'bypass or short circuit the TUC General Council' and he suggested that 'the real object of the SMA really is *to learn something of industrial health problems* by pretending that they can exchange some useful specialised knowledge'.[73] In 1947, the SMA did indeed decide to bypass the TUC when it made arrangements for a conference on the administration of the new health services, choosing to approach individual unions directly. The TUC wrote to all health workers' associations individually to request specifically that they did not attend the event, explaining when the SMA complained that the TUC could not endorse any attempt to shape health policy which sought to bypass existing machinery.[74] Subsequently, unions and trades councils who wrote to the TUC asking for more information on the SMA were informed that it was not a trade union and to allow it to use trade union mechanisms to formulate and promote its policy was to allow 'the tail to wag the dog'.[75] 'I know of no service which this Association can give to your branches which these branches cannot already obtain through your own Council, their own unions and the TUC', the Swansea and District Trades Council was informed in a typical response.[76]

Some of the antagonism which characterised the relationship between the SMA and the TUC may well have stemmed from confusion as to the role and function of the TUC. This issue was at the forefront of a dispute between the two organisations in 1951 when Tewson complained, via Morgan Philips, the Secretary of the Labour Party, that the SMA had infringed on the TUC's work by issuing circulars and questionnaires to trade unions and trades councils on the subject of industrial health, circumventing official TUC policy. The SMA asserted that the questionnaire aided worthwhile research which in no way interfered with the 'legitimate field of trade union work'.[77] Grumbling to Morgan Philips, Tewson claimed that:

> The SMA, having issued a six-page circular and questionnaire dealing with a wide range of health matters, every one of which is under continuous review by the General Council and its affiliated unions, propose – (a) to issue a report based on this Enquiry; (b) to convene Conferences, following which recommendations based on the report are to be made; and (c) to organise a campaign for their implementation.

'Dr Stark Murray', he continued, 'must have a widely different conception from ourselves as to what constitutes interference, or alternatively "the legitimate field of trade union work".'[78] Morgan Philips suggested a meeting between the two bodies to smooth out difficulties, but this did not appear to be what the TUC had in mind. 'I think it should be made clear to the Health Committee that in our view the SMA are usurping the functions of both the Labour Party and the TUC', the TUC's Social Insurance Secretary advised Tewson.[79]

In this era, the TUC devoted its energies to developing a closer relationship with the BMA, despite it being abundantly apparent that the SMA's stance on health issues reflected far more closely the labour perspective on health care policy, the SMA's affiliation to the Labour Party, and despite the repeated attempts made by the SMA to engage with the TUC. The SMA was perceived as a threat for a number of reasons. The TUC felt that its relationship with the BMA, forged through the Llanelly crisis, was the most influential route to influence and collaborate with the medical profession and was unwilling to jeopardise this relationship by according similar privileges to the SMA. The TUC was also concerned that the SMA would exploit trade unions to extract information with which it could draw up plans for a health service that claimed to represent the interests of the workers, pre-empting any official TUC statement on the issue, and stealing the TUC's thunder. A fundamental problem here was the agonisingly slow pace at which the TUC was able to formulate its health policy, through the assent of affiliated unions and in tandem with the Labour Party. The SMA's radical stance, including the admittance of communists to its ranks, was another source of anxiety for the TUC which was attempting to court support from across the moderate political and professional spectrum in this era and did not intend to advocate plans for a nationalised health service which differed from the plans being drawn up by the Labour Party. Correspondingly, the SMA viewed the TUC's relationship with the BMA with concern, anxious that the BMA would moderate the TUC's health policy.[80]

The limits of voluntarism: health and welfare schemes

Employers' motives for inaugurating welfare provisions within their factories to create a working environment which exceeded the minimum legal standards, and workers' reception of such schemes, was much debated at the time and remains a contentious subject among historians, as the previous chapter explored. In many instances, welfare schemes were instituted specifically to attract and meet the needs of women workers, and the specifically gendered discussion of welfare facilities will be discussed in the next chapter. However, by the interwar years it had become common practice for welfare provisions to be designed for both male and female employees. In his 1924 work, *The Principles of Industrial Welfare*, John Lee claimed that around 1,300 British firms had adopted a system of welfare work.[81] Lee argued that welfare work was not antithetical to trade unionism; its function, he claimed, was to enrich workers' minds and counter monotony. Trade unions, he asserted, adhered too rigidly to materialist ideals.

In 1932 the TUC produced a report analysing the impact of welfare schemes on trade unionism, drawing on material from industrial welfare organisations and surveys completed by affiliated unions.[82] The TUC's survey drew heavily on literature produced by the Industrial Welfare Society. This organisation claimed to adopt a neutral stance; nevertheless, employers cited within this literature expressing their hope that industrial welfarism would undermine trade unionism by improving the trust between workers and employers. The Managing Director of Mavor and Coulson, for example, stated at the 1926 Industrial Welfare Society conference that 'with better leadership there would be fewer followers for those who incite hostility and preach that employers are the natural enemies of the workers'.

The majority of trade unions which responded to the TUC survey – 14 out of a total of 24 – stated either that welfare schemes had not had a detrimental effect on trade union organisation, or that they had no knowledge of welfare schemes where their members were employed. Of the remaining ten trade unions, the Leeds District of the National Union of General and Municipal Workers was of the opinion that industrial welfare was essentially just frills which sought to distract workers from their real needs, noting that while employers sought to gain the upper hand over trade unions by providing their employees with social activities, 'workers find they are quite alone in matters of workmen's compensation unless they have trade union support'. The South Wales District of the same union was suspicious of the emphasis placed upon creating a bond between employer and employee and breaking down class antagonism, fearing that welfare schemes, and in particular sporting facilities, created too much 'good fellowship' between the workers and the manager: 'A sports organisation, with its natural basis of 'playing cricket' and general good fellowship, gives the employer a splendid foundation for seducing the workers from their unionist ideals by the insinuation that since employer and employed are all good fellows together, their real interests are exactly similar.' Although improvements in health provisions at small firms where conditions were still poor would be welcomed, the District expressed concern that highly evolved welfare schemes would retard progress towards a unified national health service. The Midlands branch critiqued the development of works committees, lauded by industrial welfarists as a means to encourage workers to participate in the management of the workplace, arguing that they could lead to a decline in union membership as the matters dealt with by these committees usurped what were formerly seen as trade union matters and that consequently some workers felt therefore that there was less reason to join the union.

In its report, the TUC attempted to draw a balance between denigrating the impact of welfare on trade unionism and supporting the improvements in factory environments often advocated by industrial welfare supervisors, such as improved lighting, seating and ventilation. It expressed concern that educative and welfare facilities organised by employers would encourage workers to view the firm and not the union as the organisation which looked after their interests, though it noted that this danger was avoided if the sports organisation was financed and run by the workers. The TUC viewed schemes instituted to improve thrift and cooperation with the most suspicion, arguing that 'the bait of economic security is used to turn the workers from the unions and destroy that basis of class solidarity in which alone rests any hope of future betterment'. The more financial interest workers were induced to have in a company, the TUC argued, the greater their reluctance to jeopardise the profitability of the company and the greater their loyalty. Ultimately, the TUC felt that voluntary initiatives militated against nationwide improvements. The ideal to be aimed at was a comprehensive state pension and insurance scheme; employer schemes limited workers to employment within one establishment and militated against trade unionism. Friendly societies financed by employers, some of which stipulated that members must not be trade union members, had a damaging impact on trade unions which had already been financially hit by the Depression and had had to reduce benefits. The timing of this report was probably inspired at least in part by the difficulties experienced by the TUC as a result of the Depression. Financial schemes may have been targeted as an area of particular concern by the TUC at this stage because the unstable economic environment and high levels of unemployment would have made financial security for the worker an even greater concern than usual, which employers, as Melling argued,[83] were keen to exploit.

One way the TUC and individual trade unions could stage a challenge to the health and welfare facilities provided by employers and regain the initiative was to establish its own facilities. By the mid-1930s, the TUC seriously considered propositions put to it by several psychiatrists who argued that treatment and care for workers experiencing mental health problems as a result of their work or an accident at work should be provided by workers for workers.[84] The correspondence and memoranda on this subject illustrate that health care professionals had begun to view the TUC as an organisation which was legitimately interested in the provision of health care and capable of funding and providing its own services. The proposals brought before the TUC critiqued the current provision of health care services, arguing that middle-class assumptions and illness categorisations were imposed upon sick workers while the social conditions which had brought about the disorder were disregarded.

In 1935, Dr Rosenfield, a psychologist at the Manor House Hospital contacted the TUC proposing that it establish and run mental hygiene clinics for industrial workers.[85] 'It will be both timely and politically expedient that workers should run their own health organisations', Rosenfield wrote to Smyth.[86] 'To expect employers to take action to improve workers' conditions is utopian.' Rosenfield castigated employers for using welfare as an aid to boost profits, while displaying little real interest in the health of workers. He argued that working conditions, hours of work and methods of remuneration could cause mental illnesses, while employers asserted that mental illness was caused by personal or family factors. Rosenfield also attacked the class bias of psychiatry and the failure of many psychiatrists to examine the patient's environment.[87]

It is perhaps indicative of the moderate path pursued by the TUC in the interwar years as it courted the government, the BMA and employers that concerns were expressed about the left wing political bias of the proposed clinics. Ultimately, however, the proposal appears to have floundered on the question of (financial) responsibility. In a letter to Morgan, Smyth queried why workers should pay for clinics rather than campaigning for the State to do so, given that it was employers who were responsible for workers' sickness; these concerns led the TUC to reject Rosenfield's plans.[88]

J. R. Rees, the Medical Director of the Institute of Medical Psychology at the Tavistock Clinic, hoped to entice trade unions, employers and insurance companies to fund the Institute's work on economic not philanthropic grounds. Returning to the language of the scientific study of industry first deployed by the HMWC, Rees argued that psychoneuroses were responsible for one-third of all sickness and cost businesses 10 million weeks of working time each year through absenteeism and inefficiency. He cited the assertions of Dr Lockhart that in some industries as much as 60 per cent of time lost through sickness could be attributed to psychoneurotic illness. Businesses, he argued, should contribute 'largely as a business proposition', claiming that 52 per cent of the cases treated by the Institute were industrial workers.[89] It appears that neither employers nor the TUC were attracted to Rees's project however: a memorandum produced for the National Association of Trade Union Approved Societies by J. W. Yerrell in October 1937 noted that the Tavistock Clinic was closed for ten months of the year owing to lack of funds, which suggests that Ree's appeals to employers may have been met with the same indifference which the IHES had faced. Yerrell, meanwhile, described the Clinic's approach as 'the traditional method of private enterprise and private exploitation in the treatment of disease'.[90] While the TUC dismissed the work of the Tavistock Clinic, Dr E. N. Snowden's proposals for the rehabilitation of workmen after accidents met with a more favourable response, although ultimately no services. Snowden had worked both with cases of shell shock and in the psychotherapy department at St Bart's, and Morgan noted approvingly that Snowden tended to favour the worker.[91]

While trade unions may have been wary of providing rehabilitation services which suggested that the health problems of their members had a psychological basis, they enthusiastically supported orthopaedic treatment.[92] The Manor House Labour Hospital, which specialised in providing treatment to victims of industrial accidents and industrial diseases amenable to physiotherapy, enjoyed long-lasting support from both the TUC and individual trade unions.[93] Established during the First World War to treat wounded soldiers, the hospital came under the control of the newly constituted Industrial Orthopaedic Society, a friendly society which offered free treatment to working men who subscribed 1d per week.[94] In its publicity, it actively targeted trade unionists, asserting that support for the hospital was a natural extension of trade unionism. One leaflet, entitled 'The First Labour Hospital: Workers of the World Unite', asked its readers, 'Why do you belong to a trade union?'

> Because you believe that by combined effort you can accomplish that which individually would be impossible, your power as an individual is small – so small as to be almost impotent. On the other hand, your collective power is invincible.[95]

The hospital received a number of grants from individual trade unions and the TUC to repair damage from air raids, expand the premises and update equipment. The TUC, the Amalgamated Engineering Union and the Electrical Engineering Union all endowed beds within the hospital in the 1940s for between £1,000 and £2,500.[96] Indeed the TUC contemplated establishing and maintaining an entire ward for industrial diseases, envisaging that this would act as 'a lasting memorial to the humanitarian side of trade unionism ... a pioneer movement worthy of the highest praise and commendation'. The appeal was partly premised on need – no facilities of this nature existed at the time of writing – but the memorandum dwelt more on the impact it would have on the status and public image of the TUC. 'No industrialist or philanthropist has yet endorsed even a bed for industrial disease', the memorandum noted. 'The trade union movement could blaze the trail.'[97] Ultimately, the estimated construction cost of £10,000 and annual running costs, also estimated at £10,000, led the TUC to endow a one-bed ward at a cost of £2,500 instead.

Trade union support for the hospital continued after the inauguration of the NHS. In 1949, the Transport and General Workers' Union donated £10,000 for the establishment of an industrial diseases medical institute to commemorate colleagues who had died during the war. Donations of £1,000 were also received in that year from the Amalgamated Engineering Union, National Union of General and Municipal Workers and the National Union of Railwaymen. Indeed, the hospital continued to function with the support of trade unions until 1999. The extensive financial support afforded to the hospital by trade unions belies the argument that trade unions were disinterested in health issues; what it does suggest is that there was a greater willingness to support projects which sought to remedy visible physical damage caused by injury or disease, rather than invest in preventive or mental health services. Listing the problems inherent in a worker-run health education service in 1938, Morgan noted 'spending money on health work which brings in no apparent immediate return'.[98]

This thesis is borne out in the example of trade union convalescent homes. A file of correspondence between the TUC and affiliated unions on the subject of convalescent homes demonstrates that interest in convalescent provision began to develop from the late 1920s and that it was individual unions rather than the TUC which played a pioneering role. Newspaper articles and pamphlets described a range of convalescent homes instituted by individual unions for their members, such as the Joseph Cross Memorial Convalescent Home, run by the Amalgamated Weavers' Association, praised in the *Preston Guardian* for its 'homely touch'.[99] However, a letter received by the TUC from the Society for Cultural Relations between the Peoples of the British Commonwealth and the USSR enquiring about provision for industrial invalids forced the TUC to acknowledge that no such information was collated. Subsequently, Citrine wrote to all affiliated unions asking what services individual unions had instituted.[100] The responses received suggest that convalescent provision had become an essential service for many unions, and that individual unions had responded to this in different ways. Some unions paid annual donations to hospitals which entitled them to supply their members with letters for convalescent and hospital treatment. Other provisions were more elaborate: a pamphlet detailing convalescent provision for the National Society of Operative Printers and Assistants described not only the convalescent home and the home for aged members, but the planned sanatorium for tuberculosis sufferers. Trade unions may have seen the provision of such health care facilities as a means to entice members' loyalty at a time when trade union power was weakened, especially as the gravest threat believed to be posed to trade unions by industrial welfare in this period were the financial insurance schemes which offered to protect workers in times of illness.

The example of the Manor House Hospital and convalescent care testifies to the willingness of trade unions and the TUC to invest in remedial care services. The relationship between the TUC and the IHES provides a rare example of the TUC's willingness to collaborate with another organisation in the pursuit of health promotion in the 1920s, prior to the appointment of the TUC's first Medical Adviser Sir Thomas Legge in 1930.[101] Following Legge's appointment, the relationship between the IHES and the TUC altered radically. One of Legge's first tasks upon his appointment was to investigate the workings of the IHES and it appears that the issues raised by the TUC's collaboration with the IHES may have been a factor in the TUC's decision to appoint a medical adviser. The relationship between the two organisations also illustrated another vexing issue which was to become a major consideration in TUC policy in the 1930s: who exactly should take responsibility for health.

The IHES was founded in Scotland in 1924 as the Industrial Educational Council, with the objective of 'educating industrial workers on the subject of sicknesses and diseases to which they are liable in the nature of their occupations'.[102] Primarily, the organisation arranged for doctors to give talks to workers in various industries, such talks including in 1925 'General Health Talk', 'Occupational Sicknesses and Diseases' and more specific subjects such as 'Eye Strain, Chest Contraction, Tuberculosis', which was given to the Prudential Staff and Amalgamated Union of Insurance Workers. The IHES approached the TUC for endorsement in April 1925. This was officially granted in January 1926, by which time the organisation had already provided lectures to various groups of workers, including mill and factory operatives and textile workers.

The IHES canvassed for support among employers with promotional literature which stressed the amount of work lost through sickness and disablement that could be prevented if the workers had the necessary knowledge. Attempting simultaneously to court the TUC and the BEC, the IHES sought to eschew a political stance. In a statement given to the TUC's General Council, the Secretary of the IHES Mr Mackenzie noted that he had frequently heard it stated at trades and labour council meetings that wages and hours of work had a great effect on the health of a person. However, Mackenzie argued that 'it is the policy of the Society to keep out of controversial subjects such as politics and trade union matters', seeing such issues as being the rightful business of trade unions.[103] He stated also that employers were not allowed to influence the policy of the society. However, the IHES tended to popularise its work using the language of industrial efficiency which had characterised the reports of the HMWC. 'There can be little doubt about the benefit to workers, to industry, and to the community at large', an article from the *Scotsman* stated. 'Apart from the personal benefit which workers derive...from the health lectures of the Industrial Education Council, their work is well worthy of support as a valuable aid to increased industrial efficiency.'[104]

Throughout the late 1920s, the IHES appeared to be successful in attracting unconditional support from the TUC. An article purportedly written by TUC President Ben Tillett entitled 'The Health of the Worker', which full-heartedly endorsed the work of the IHES, was distributed through trade union publications. Claims within the article that 'the TUC is much concerned with the health of the industrial and other workers' were slightly compromised by the fact that the article was actually written by the IHES and that Tillett simply endorsed the piece.[105] Further articles sent by the IHES for inclusion in trade union journals stressed the responsibility of the worker for his own health: 'For his physical defects the worker is in some measure himself responsible. He is often careless and indifferent regarding the things on which physical fitness most depends: his rest, his exercise, his food and drink, to say nothing of the general hygiene of his home.'[106] Around 450 talks were arranged by the IHES in 1928 and in 1929 the TUC decided to increase its annual £50 subscription to the IHES to £100. Pursuing his advantage, Mackenzie asked for subscriptions from individual unions, arguing that unions had hitherto expended large sums of money in sickness benefits but invested little in preventive health care.[107]

While the TUC complied with the IHES's requests for endorsement and financial support, the BEC were proving harder to entice. Venting frustration, articles promoting the IHES noted that it had been the employees' side which had first recognised the need for prevention and education, while the employers had failed to respond.[108] However, while IHES articles heaped praise on the TUC for pioneering preventative health care for their members in addition to more traditional trade union functions, it is clear that the initiative lay with the IHES and that the TUC was a passive partner in the relationship.

This was to change radically in 1930 with the appointment of the TUC's first Medical Adviser, Sir Thomas Legge, who attended the annual meeting of the IHES and expressed a number of misgivings.[109] He criticised the 'farcical' organisation of the Society, noting that of the 61 Executive Committee members and 246 Sub-Committee members listed in the pages of the annual report, only 20 showed up to the meeting, and that three of those came late or left early. Legge also attacked the Society for allowing advertising revenue to distort the information it presented to workers, citing the example of a flyer for a health talk which on the reverse carried an article entitled 'Why Should We Eat More Fish?' by a Dr D. H. Geffen. It later transpired that the Society had received £200 for printing this advertisement/article on the reverse of their flyers, which claimed that meat consumption put a strain on the kidneys.[110] Most damning of all, Legge queried the accuracy of the information given by the IHES to workers. As a recognised expert on the subject of industrial lead poisoning, he confidently critiqued articles written by the IHES on the subject, asserting that contrary to the information provided by the Society, around half of all lead workers were affected and that the sickness could come on rapidly. He also questioned the expert knowledge of the doctors providing lectures for the Society on the subjects of occupational sickness and diseases, claiming that only four out of 44 had the requisite skills. Finally, Legge queried the value of educating workers about how to keep fit if employers did not become involved. 'I came away from the meeting feeling that the organisation was thoroughly bad', wrote Legge, 'yet it has attained, in a way, a big backing.'

Legge's biting assessment of the IHES appears to have given the necessary confidence to other, non-medically trained TUC staff, to critique the work of the Society. Herbert Tracey, who worked in the publicity department, contacted Legge and subsequently Citrine with his concerns about the quality of the articles submitted by the IHES for publication in trade union journals such as the *Industrial Review* (Legge noted that in one instance the word 'silicosis' had been misspelled five times in an article on that subject).[111] It appears that concerns had arisen about the utility of the IHES before Legge's appointment and that one of the first things he had been asked to do was to examine the society.

As Legge had resigned his post as Medical Inspector of Factories in 1926 over the government's failure to ratify international conventions banning the use of lead paint inside buildings, it is unsurprising that his attacks focused on the Society's coverage of lead poisoning. Legge felt that certifying factory surgeons would be far better qualified to give talks on the subject of industrial disease and occupational health than the medical officers of health favoured by the IHES, an early example of the medico-political rivalry between practitioners employed by the Ministry of Health and those employed by the Factory Department which would come to dominate debates on the future of industrial health services after the Second World War.[112] Legge suggested that medical officers of health should restrict their lectures to subjects such as general health and nutrition. He also challenged the memorandum issued by the IHES to speakers which asked them to 'refrain from answering questions by the workers on Political, Social or Industrial matters, which are the concern of employers and employees, such as hours, wages, and conditions of labour'.[113] At an executive meeting of the IHES held in July 1930, at which only seven people were present and 'a desire was frequently expressed to get on with business so as to have lunch', Legge claimed the memoranda prevented speakers from informing workers how real protection could only be secured first by the employers doing everything necessary. He cited an example of a lecture in which workers had been advised to wet rub down paint, but 'there was nothing in it to suggest that the employers should do anything...not a word on the substitution of harmless paint'. Legge noted with concern that the Association of Lead Manufacturers had subscribed £20 to the IHES with an additional £1 for every lecture given to workers on the subject of lead poisoning. Noting that the Association had opposed the draft White Lead Convention, Legge argued that in all likelihood the Association was funding the IHES to protect the status quo. He castigated the ignorance of the IHES's Secretary Mr Mackenzie 'who does not seem to realise how impossible it is to address workmen on Silicosis or Dermatitis or Lead Poisoning or Skin Cancer without often having to refer to social and industrial matters, and occasionally to political'.[114]

A letter sent by Legge to the Executive Committee of the IHES reiterated his twofold objections to their work, criticising both the ignorance of the lecturers 'who have been content to invade perhaps the most specialised branch of medicine' after reading a leaflet on the subject, and the belief of the Society that workers should be educated so that they could protect themselves from illness, which ran counter to his axioms: 'until the employer has done everything...the workman can do next to nothing to protect himself'. Legge described a lecture he had attended on lead poisoning in which the lecturer – either through ignorance or a desire not to trespass on social, political or industrial matters – had been unable to answer the question 'Would it not be a good thing to abolish white lead altogether?' Perhaps hoping to avoid a confrontation with Legge, the lecturer deferred the questions asked by workers at the lecture to Legge, who subsequently received a written rebuke from the treasurer of the IHES for touching upon social, political and industrial matters.[115]

To placate the TUC, the IHES agreed that lecturers were free to make reference to any regulations, Acts or rights and duties of employers, workers and the government. Commercial concerns were no longer to be approached for grants of money and an effort was to be made to secure more suitable speakers for talks on occupational diseases.[116] Throughout 1931 and 1932, the IHES expanded its range of lectures, providing more talks on issues of health and minor ill-health including the health of young workers and women workers, nerves, anaemia, fatigue, constipation and varicose veins. The Society also began to build up a collection of films on subjects including tuberculosis, diphtheria, clean milk, influenza, dental health and the benefits of sunshine. While the TUC continued to endorse the work of the IHES, its endorsement was more explicit about the TUC's own work in health promotion and was not without reservations. A letter written by Citrine at Mackenzie's request stated that 'the worker's health is more important to him than anything else...we give great attention to that side of the problem here and we are glad to cooperate with other bodies having the same end in view. These talks, *provided that the lecturer understands his subject*, are an ideal way of demonstrating the dangers'.[117]

Following the death of Sir Thomas Legge, Mackenzie wrote to the newly appointed TUC Medical Adviser Hyacinth Morgan congratulating him on his appointment. 'The late Sir Thomas Legge was closely associated in his work with this Society and I trust the same happy relationship will exist between the Society and yourself.'[118] If Mackenzie had hoped for a more harmonious relationship with the TUC he was to be disappointed: Morgan's appointment was to usher in further scrutiny of the Society. Memoranda produced by Morgan on the IHES reiterated Legge's anxiety that the doctors giving the lectures often lacked the necessary expertise, and he voiced his concerns that the ideas expressed might be contrary to the official stance of the TUC. The memoranda critiqued the undemocratic nature of the Society, depicting it as a one man show. Although the vast bulk of the Society's funding came from labour sources, including the TUC and the Miners Welfare Fund, Morgan noted that trade unions had no control over the policy, programme or work of the Society as neither the Finance Committee nor the Executive Committee contained any trade union representatives. Morgan argued that better work could be done if the work was organised and financed by the TUC's General Council.[119] The TUC's annual grant to the IHES was renewed in 1933 but it was agreed that Morgan would report on the ramifications and efficacy of the Society's work. Smyth contacted those present at IHES talks for their impressions, and received one response complaining that the lecturers 'stressed very strongly indeed what the workman should do but said nothing about what the employers should do'.[120] Mr W. M. Emerson of the Lowestoft Unemployed Association wrote to the TUC in 1935 complaining that the IHES had been unable to find a doctor to give a health talk. 'My view', Emerson wrote, 'is that the local Doctors are suffering themselves from the malady of "Toryism"'.[121]

Anxious that the TUC would withdraw support, the IHES offered the TUC first three then six representatives on the Executive Committee of the Society in August 1935, which the TUC accepted, though it rejected the IHES's application for an increase in its annual grant. It appears that the TUC became increasingly impatient with the inability of the IHES to attract money from employers. Venting his resentment that trade unions were funding something that benefited employers and insurance companies who did not contribute, Smyth wrote to Citrine, 'I really think that the employers and insurance companies should be made to face up to their liabilities in this matter, or alternatively that the Government should develop its own health services to cover industrial diseases'.[122] However, accepting liability for workers' ill-health appears to have been the last thing that most employers wanted to do, and indeed this seems to have been one of the primary reasons why the BEC refused to fund or endorse the work of the IHES.[123] Writing to J. B. Forbes Watson, the Secretary of the Confederation, Adam Nimmo expressed deep reservations about the work of the IHES:

It is just the kind of work it is very difficult to keep upon sound industrial lines as there is a constant tendency arising to use it for political purposes... it tends to concentrate upon industrial disease to such an extent as to provoke a public enquiry followed by legislation. It does not, of course, do for one to be too much afraid of these considerations provided one is satisfied that a really satisfactory end in the interests of industry is being pursued, but I am afraid that they cannot be altogether ignored by those engaged in industry who are being constantly worried by industrial legislation.[124]

Mackenzie finally appeared to acknowledge that funding would not be forthcoming from employer organisations, writing to Smyth 'seemingly the average employer is unwilling to admit that the processes in which he is engaged are in any way harmful to the health of the workers'.[125]

While frustrated that the IHES had failed to secure funding from employers and dubious about the utility of the lectures it provided, TUC officials such as Smyth and Morgan acknowledged the importance of health education. In 1935, a series of memoranda were exchanged between TUC personnel, mooting the possibility of establishing a health education service financed and run by trade unions. Citrine described a meeting in which Ernest Bevin had raised the issue of health education among trade unionists. Bevin was a member of the Industrial Health Research Board, and was unimpressed with the progress it was making.

He thought that we only dealt with it [i.e. health] now as a hazard, that is to say, we dealt specifically with industrial diseases and with accidents and workmen's compensation, and also preventative legislation, through the Factories Act, and such like measures. We should move from this stage to a continuous examination of building up health education among the people whom we represent.[126]

Bevin urged the constitution of an advisory committee on health education which could include medical practitioners. Such a committee could provide unions with material regarding health, social insurance and accident prevention and perhaps arrange a series of talks via the BBC.

Morgan responded by outlining two different schemes which he believed the TUC could embark upon: either a programme of general health education, or an extension of the industrial health work already undertaken by the TUC.[127] He felt that the TUC lacked sufficient resources to undertake research itself. Morgan acknowledged

Bevin's argument, noting that the TUC currently concerned itself only with the threats posed to health by industrial hazards, and he wondered whether the General Council was contemplating a more generalised scheme, encompassing public health, physical and mental health, personal hygiene and the effects of the environment on health. In line with the ethos of the interwar health movement, discussed in Chapter 2, Morgan argued that a health education programme on these lines would require knowledge of economic and social issues such as housing, education, playgrounds and health care institutions. Indeed, Morgan argued that a programme of general health education would be a scheme of national importance, opening up the issue of a national medical service. Challenging the exclusive right of medical professionals to debate and determine health policy, Morgan suggested that various voluntary organisations could be consulted, but felt that membership should be largely drawn from within the trade union movement and that representatives of industrial firms should be excluded as 'their view and outlook as to ultimate effect is entirely different from that of the trade union movement'. Pamphlets, lectures, conferences, films and radio broadcasts on specific diseases, general health, industrial hazards and psychology were all suggested by Morgan as possible forms of publicity. The proposed committee, Morgan suggested, could examine subjects such as local authorities and medical appointments; sanitation; health propaganda and trade unionists; medical education; hospital policy and accommodation and trade unionism; psychological treatment; the medical profession and information as to trade unionism and medical research.

Warming to his subject, Morgan authored a further memorandum in June, arguing that the recent affiliation of the Medical Practitioners' Union, the Association of Local Government Officers and Women Sanitary Inspectors to the TUC had stimulated interest in general health issues and made a general health committee more desirable. 'Trade unionists are not only interested in health questions specifically pertaining to their work or occupation', argued Morgan, 'but also on larger national or municipal health issues which affect them partly as citizens and partly as workers.' Morgan pointed to the growing involvement of other organisations in issues of public health, such as the recent reports on nutrition by the BMA and the Ministry of Health, arguing that the TUC also should formulate a general health policy on such issues. He argued that a TUC committee could play an important function, pointing to 'the necessity for scientific opinion on social problems to be tempered by the lay trade union mind and experience'.[128]

The decision not to proceed with this particular scheme is perhaps most accurately understood as a result of the TUC's adoption of a new strategy rather than a disinterest in health, though the question of financing such a scheme no doubt played a part.[129] The general expansion of TUC interest in health improvement was, for example, reflected in its campaign to secure paid holidays for workers, in which the TUC argued that 12 days' annual holiday with pay benefited health, improved the value of leisure time and reduced absenteeism.[130] Ultimately, the TUC's disappointment with the work of the IHES appears to have put a dampener on future health education initiatives, which were seen as too politically problematic, potentially impeding state initiative. By the late 1930s it expressed considerable anger that workers were paying for services to combat the harm done to workers' health by employers, while employers were abrogating all moral and financial responsibility. A memorandum noted the TUC position:

> It was most unfair that the workers should be continually asked to pay money out of their own pockets to finance propaganda in regard to the prevention of industrial diseases which was obviously a matter which the employer was responsible for... If the employers were not going to help... then we might have to reconsider our position and press for this work to be done by the Government and paid for out of taxation.[131]

In October 1938, the General Council of the TUC resolved to withdraw both its representatives and funding from the IHES. The Council of Health Education, whose relationship with the TUC during the Second World War was discussed in the first chapter, faired little better after the cessation of hostilities. At a 1948 meeting of an ad hoc committee on industrial health, C. R. Dale, who by this stage had replaced Smyth as the TUC's Social Insurance Secretary, queried the advisability of further coordination of health education. Pointing to the work being undertaken by the Industrial Health Advisory Committee of the Ministry of Labour and National Service, which was by this stage exploring the question of an industrial health service, Dale questioned the advisability of starting coordination in this area 'de novo'.[132] TUC intransigence prompted the Council to rename its committee 'General Health Education at the Place of Work'. This new name came with a new remit which would 'make clear that the Committee was concerned only with general health education in the principles of healthy living and not with education covering industrial health hazards'.[133] Even then, the TUC remained ambivalent. 'It was difficult to see how the proposed education could be divorced from industrial health problems', the TUC's Social Insurance Committee noted.[134] Likewise, the assertion of the Council's Medical Adviser Dr Sutherland that the worker could be thought of in this context 'merely as a citizen and possibly as a parent, and not as a worker' did not convince the TUC to embrace the scheme with enthusiasm.[135]

From voluntarism to legislative solutions: trade unions and the state

During the course of the First World War, the State had been in the vanguard of industrial health policy. In the interwar years, however, much of its impetus dissipated and the under-resourced Factory Department attempted to quell discontent and foster improvements in working conditions by appealing to cooperation between employers and workers, advocating voluntary improvements. With no new substantial legislation dictating minimum health and welfare standards within factories until the 1937 Factory Act, aside from the 1924 Workmen's Compensation Act which required the provision of first aid facilities in all workplaces, the Inspectorate saw initiatives such as the Home Office Industrial Museum, discussed in the previous chapter, as the most effective means to raise standards.[136]

'At no time in the history of factory legislation have the relations between the Inspectorate and the two chief partners in industrial progress – employers and workers – been closer or more friendly', proclaimed the Inspectorate in 1937.[137] This rhetoric of friendly cooperation helped to disguise the fact that this was not an era to enforce improvements via legislation. Many employer organisations were particularly resistant to imposing standards via welfare regulations, often citing the dire economic situation and the costs involved. The trend toward voluntary standards, which not all employers adopted, led to widespread discrepancy in working standards. In 1930, for example, the Inspectorate noted that the draft Safety in Factories Order remained in abeyance following assurances from employers' associations that they would take up the matter of safety organisation with their members.[138]

The 1937 Factory Act legislated for new standards within workplaces, regulating facilities which had previously come under the ambit of voluntary welfare provisions, lowering maximum working hours, reworking the system of certifying surgeons and enabling the Home Office to order medical supervision in instances where it was feared that working practices could damage health.[139] However, this Act was only passed after 15 years' campaigning for new legislation. A first draft of the Bill was produced in 1923 and was introduced to Parliament in 1924 but did not get a second reading. It was promised in the King's speech of 1924 but was not introduced. A revised Bill was introduced in 1926 but only for the purposes of discussion.[140] 'In view of the repeated attempts that have been and are being made by employers' associations and similar bodies to secure the dropping, postponement, or toning down of the Factory Bill', a 1927 TUC memorandum noted, 'we think it necessary to restate with emphasis the reasons why it should be proceeded with immediately, and in a much stronger form'.[141] This statement was written a decade before the legislation was passed. In view of other commitments, the government decided to postpone the Bill in 1928. It was again promised by the Labour government in 1930 but once more was crowded out by other legislation. Citrine expressed his exasperation at a deputation with the Secretary of State John Simon in 1936, comparing the Factory Bill to a 'bashful young debutante' who had been waiting for 35 years to be introduced. In response, Simon suggested that the Factory Bill was 'not a slight little thing who could squeeze in anywhere; it is a weighty and long-winded female, and it takes some time'.[142]

Until the passing of the 1937 Act, workers had to rely on the existing provisions of the 1901 Factories and Workshops Act. Although the numbers of factory inspectors increased from 138 in 1901 to 290 in 1937, when it was announced that the number was to be expanded to 332, there was a limit on how much an Inspectorate of this size could achieve given the vast number of factories and workshops which it had to inspect, let alone the limitations placed upon the tiny numbers of medical inspectors. These shortcomings were recognised by Legge, though he still made suggestions to the TUC on how best to utilise the expertise of the Inspectorate through informed complaints.[143] By 1937, lectures were being given by inspectors to local trades councils at the suggestion of the TUC to explain the law to workers.[144]

Throughout the interwar years, debate rumbled on between trade unions, employers, medical organisations and political parties about possible national health service schemes and where these should be located. Several controversial questions fractured possible accord: were employers, workers or the State ultimately responsible for workers' health; should health care be provided within or outside the workplace and to what extent did work affect health? These discussions could be informed by international examples. In his capacity as Chief Medical Inspector of Factories in 1920, Thomas Legge noted somewhat enviously that in America, where no system of National Health Insurance was in place, a system of employer-provided medical supervision had arisen in the large factories employing over 2,000, both to aid the medical selection of workers and to keep them in good health. Legge also noted that Workmen's Compensation Acts in several states required medical and surgical treatment as well as sickness and disablement benefit. Such services could include surgeries, dental and ophthalmic departments, though little or nothing was done in smaller factories. This growing market for industrial medicine fuelled the rise of training courses, which included topics such as factory sanitation, preventative health, educational propaganda, home nursing of sick employees and the improvement of home and community conditions.[145] Ten years later, the same unfavourable comparison lay at the heart of a memorandum produced for the TUC by Legge, in which he compared the 50 full-time officers employed within British factories with developments in America.[146]

A memorandum from 1936 noted that TUC policy on health provisions fell into two categories: '(1) Our ultimate ideal, and (2) Those improvements and developments in the existing medical services which may be considered expedient and practicable in the near future'.[147] While the TUC ultimately wished to see a state medical service under government control which was available to all and provided free of charge, it felt that in current circumstances an extension of National Health Insurance scheme would have most chance of success. This entailed the creation of a worker-centred nationwide public health service, which aimed to keep the working population healthy but which brought in provisions that were currently run separately, such as hospital services, child welfare, maternity, treatment of dependents and specialist services such as dental treatment, even if those health care provisions were based largely outside the workplace.

Of all the TUC personnel in this era, Legge appears to have been the most opposed to an extension of the National Health Insurance provisions. Fundamentally, Legge believed that workers' ill-health originated in the workplace and thought it ludicrous that health services provided for workers were 'quite extrinsic to either the factory or the workshop; the reason being that workpeople in the factory have to wait until they are ill and then stream to their panel doctor, whereas there is not even a trickle from the panel doctors into the factory to try and prevent sickness claims arising which must ultimately fall on the National Health Insurance funds'.[148] Undoubtedly, Legge's position on this issue had been shaped by his 28 years' service in the Factory Inspectorate. As we have seen in the example of the IHES, Legge believed that the Ministry of Health and its services lacked the expertise and setting to deal with industrial health issues, and wanted to see the matter developed by the Factory Department of the Home Office. While a great extension of the Factory Department's work in health matters was out of the question in the interwar years, it once more became a seriously considered option after the Second World War when the establishment of a national industrial health service was under consideration.[149]

The National Health Insurance provisions were also criticised by the TUC on the grounds that they failed to prevent illness. In a deputation to the Ministry of Health, it complained that:

The original Act, [the National Health Insurance Act] as you know, was designed for the purpose of preventing illness but in our view it has deteriorated into nothing more than an instrument for the payment of cash and the supply of medicine to people who fall ill. Claims were made for this Act at its inception that the purpose was to regard the health of the people as the chief vested interest of the nation... In our view exactly the opposite has taken place.[150]

The TUC delegation expressed regret that provisions under the National Health Insurance Act to hold employers responsible for excessive sickness and make them refund to the approved society or the insurance committee the extra costs incurred had been dropped because of an absence of proper statistics. The TUC complained that approved societies, with the exception of a few approved by the trade unions, were neither occupational nor geographical, and that consequently their potential to collate a wealth of statistical information on the health of the people which could be utilised to guide preventative measures was unfulfilled. The TUC deputation argued, in contrast to the line of the Senior Medical Inspector of Factories John Bridge, that 'a great number of the illnesses from which people suffer are due to occupation'.[151] However, like Bridge, it argued that statistics were required to institute effective preventative health measures, as this would demonstrate the prevalence of specific diseases in particular industries. The deputation cited the example of how one trade union approved society had recently convinced a Home Office committee that card room dust in the cotton industry caused bronchitis by the production of statistics to demonstrate the potential of the National Health Insurance Scheme, urging the Minister of Health to make the Act 'a measure for the prevention of illness and a great index of national health'.

Conclusion

Howard Collier, whose bitter indictment of trade union inertia opened this chapter, was brusquely informed by Hyacinth Morgan that 'you must allow me to take exception to your view that the efforts of the TUC in any particular direction with regard to industrial disease was inadequate. Certainly not in the time of my advisorship – and certainly not long before that.'[152] Indeed, Morgan informed one correspondent who wrote to the TUC in 1941 asking for information on the work undertaken by trade unions to protect the health of workers that their request would 'entail a considerable dissertation as especially in the last ten years (though it started long before) a considerable amount of persistent work has been done for the health, safety and welfare of the worker by the trade union movement in Britain'.[153] Conversely, while acknowledging the 'considerable expertise' acquired in the field of workmen's compensation, Paul Weindling has asserted that 'trade unions have often made pay a priority, and neglected health issues'.[154]

By centring this account of the politics of industrial health in the interwar years on the TUC and trade unions, this chapter urges a re-evaluation of the traditional assumption that trade unions pursued pay at the expense of their members' health, articulated above by Weindling. What is at stake is not so much whether the TUC sought to become involved in health issues, which I would argue is indisputably the case, but what the TUC achieved and whether it succeeded in representing the interests of its members. This, in turn, was dependent upon the interpretation of the function of the TUC and an assessment of what it could achieve within a specific political and economic climate.

Answerable to its affiliated unions and with little autonomous power, the TUC was entrusted with the task of promoting cooperation among trade unions and representing the common interests of organised labour. It could take little action without securing the consent of its affiliated unions and did not have the financial resources to establish large-scale research or treatment facilities in its own right. In this respect, individual unions and branches had far more scope for action and often did invest in health services for their members. The Llanelly dispute centred on the question of whether worker organisations were entitled to dictate and control, as opposed to simply funding, medical service for workers. Arguably, the trade unions of Llanelly lost more ground than the medical professionals, but given the paucity of medical training on the subject of industrial health in the interwar years, trade unions were justifiably cautious when medical organisations claimed that lack of medical expertise disqualified trade unions from adjudicating on health issues. The TUC's decision to appoint a medical adviser enabled the TUC to more authoritatively state its case to medical and voluntary organisations and to the government. The Llanelly dispute was also a timely reminder of the divergences which could exist between TUC policy and trade union practice. This tension between local needs and national objectives would persist into the post-war years as the TUC attempted to establish a national industrial health service, while local branches sought to make the best of what was available.

Even if it could afford to inaugurate health services in its own right, the TUC increasingly came to believe that it was not the responsibility of trade unions to pay to improve the health of workers damaged by the working conditions and practices established by employers. Much of the debate in the interwar years focused upon the respective impacts of home and factory environments on health, and consequently on whether health provisions and services should be centred in the factory or the home. Trade unions lambasted filthy, dirty, unhealthy factories and occupations, and blamed them for workers' illnesses; welfare supervisors lambasted filthy, dirty, unhealthy worker homes and blamed them for worker illness, arguing that the factory environment was healthier. Even linking very specific industrial illnesses to work processes in order to secure registration under the Workmen's Compensation Act was still a very contentious and ongoing business in this era;[155] attributing general illnesses to the working environment and work processes was yet more difficult.[156] The logical upshot of this was that first, no consensus existed as to whether services should be based in factories to protect workers from the hazards posed by work and the working environment, or based in a community setting. This also proved a crucial debate about where future national health services would be situated and what they would focus on. Secondly, debates about the respective roles of work and home environments in the production of disease essentially sought to lay blame and responsibility at the (literal) door of the employer or the employee.

Nevertheless, TUC officials, Morgan in particular, were clearly seduced by the thought of appearing as pioneers in the field of health care and found the idea of shaming employers, the government and medical associations by selflessly establishing health care provisions out of their own pockets very appealing. Such action, the TUC believed, would increase the respectability of the TUC and enhance its authority, furthering the TUC's general ambitions as well as promoting industrial health. It could be unfair to criticise the TUC for failing to do what, after all, was beyond its remit. However, the TUC was apt to interpret its function – to promote the interests of workers – very broadly, and this proved problematic for groups which shared similar objectives. Similarly, while the TUC frequently asserted (as Morgan did in his response to Collier) that a particular approach was outside the remit of the TUC's function, subsequently it could be found dabbling its toes in a very similar scheme, perpetuating the ambiguity of its function and peevishly attacking any organisations attempting to work in the same area, claiming that these groups were encroaching upon its function and the formulation of legitimate and official labour policy. Howard Collier and the SMA were hapless victims of the TUC's constantly shifting 'function' as regards health. By attempting to lay claim to a broader and more authoritative role, the TUC misled unions and political collaborators as to what it was able to do, inevitably leading to disappointment.

Health was an ambiguous concept, as the work of the IHES exemplified. The IHES argued that given the right information, workers could take responsibility for their own health, and this sentiment must have appealed to the TUC in the 1920s because it did initially fund the Society. The appeal is understandable: taking responsibility suggests agency; it suggests that workers have the power to control their environments and their health. But it also suggested that any resulting sickness was the fault of the worker, their failure to act responsibly, to eat sensibly, to live hygienically and take suitable precautions. This rhetoric was torn to shreds by Legge: what use was it, his reports essentially argued, to tell workers to eat fish if their employer insists in using lead paint which will poison them? Legge insisted that it was pointless to ask the worker to protect himself until the employer had done everything in his power to make the working environment and materials safe. This was not a mild divergence of opinion between the TUC and the IHES, but an unbridgeable chasm. Peter Bartrip drew a similar conclusion in his analysis of the regulation of occupational disease in the Edwardian era; if a healthy workplace was a matter of power and control, then it was 'invidious to place blame ... on workers who failed to wash their hands'.[157]

Legge's intervention might appear problematic. What place is there for a controversy about lead poisoning in this monograph which ostensibly explores efforts to turn the factory into site of health improvement? But this slippage between health promotion and industrial illness should not be seen as problematic: indeed I would argue that it is crucial to understand and assess TUC health policy. It points to the fact that health as it was discussed and thought about in the 1920s and 1930s was not a clear-cut concept that could be isolated by those involved and it is a reminder that health remained tied to industrial illness and responses to that: efforts to secure compensation for workers incapacitated through industrial illness and injury persisted throughout this period in tandem with endeavours to promote health. Support for this perspective can be found in work undertaken on trade union responses to industrial illness: in their exploration of trade union responses to byssinosis, Sue Bowden and Geoffrey Tweedale acknowledged that unions had concentrated on compensation but argued that unions only adopted this strategy 'because control of the workplace lay overwhelmingly with the employers and the government'. They assert that 'unions were deeply concerned about occupational health and had a relatively sophisticated understanding of the medical issues that sometimes pre-empted the experts'.[158]

It is quite remarkable that trade unions have been represented as only being concerned with securing monetary compensation for injury and disregarding the pursuit of health: the TUC was one of the most vocal critics in interwar medical politics of a system which compensated and cured, but did little to prevent illness from occurring. The attacks on the 'National Disease Insurance Act', the long campaign to ensure healthier working conditions through new factory legislation in the face of employer intransigence, efforts to alter the medical curriculum so that doctors were educated about health as well as illness and the TUC's decision to fund health education clearly testify to the TUC's work in this field. Forced to operate within a system which treated and compensated sickness and disability rather than preventing it, the TUC's work to promote workers' health developed in tandem with its extensive work to schedule diseases under the Workmen's Compensation Act. The biggest stumbling block in the way of securing improvements in health and welfare through voluntarism was the intransigence of employer organisations. One doctor argued that workers should run their own health organisations because it was futile and indeed 'utopian' to expect employers to take action to improve workers' health.[159] This was certainly an innovative argument to make, but ultimately, the unwillingness of employers to accept any responsibility for workers' health prompted the TUC to advocate the establishment of a national industrial health service, established and administered by the State. This story will be explored in Chapter 5.

4
Tailoring Provisions for Individualised Needs

Cogs in the machine: an industrial system in crisis

THE INDIVIDUAL IS APT TO BE OVERLOOKED IN A SYSTEM OF LARGE SCALE PRODUCTION.[1]

This remark, which prefaced an edited volume produced by the Institute of Welfare Workers in 1925,[2] was the mantra of industrial welfare supervisors operating in the aftermath of the First World War. They expressed concern about the impact of industrial work, the industrial environment and the relationship between the industrial environment and home life on specific groups of workers. Industry was in a state of crisis according to industrial welfare supervisors, who portrayed the relationship between employers and organised labour as one of bitter warfare, damaging both parties.

The problem, welfare supervisors argued, lay in the rapid and unplanned expansion of the factory system of production and the increasingly finite nature of the individual worker's role within the enterprise. As larger factories were constructed and the workforce of individual companies grew in size, the personal relationship between manager and worker dissolved. At the same time, the subdivision of the production process made it increasingly difficult for the individual worker to see how the specialised task they undertook contributed to the completion of the final product, robbing work of meaning and satisfaction. The worker had become a depersonalised cog in a machine. Consequently, with the earlier amity which welfare supervisors claimed had characterised relations between management and workforce in the small factory now strikingly absent, both parties looked to collaborative action as the most effective means to resolve their difficulties.

While the narrative of earlier workplace harmony had been exaggerated by welfare supervisors, tensions were undoubtedly exacerbated by the deskilling of work processes and intensification of labour in the workplace.[3] The advent of new unionism in the 1880s was a response to these developments. In place of the traditional 'craft' union, vast unions representing the interests of unskilled general workers were established, willing to adopt militant strategies to protect the short-term interests of poorly paid workers in insecure jobs. These unions believed that the existing trade union structure was reactionary and helped bolster the existing social order: it was not until 1899 that the Trades Union Congress (TUC) dropped its support of Lib-Labism, when a Congress resolution led to the establishment of the Labour Representation Committee.[4]

With a number of legal restrictions on the use of strike action lifted by the passage of the 1906 Trade Disputes Act, many unionists, inspired by syndicalist ideas, believed that coordinated strike action was the most effective weapon to deploy against employers to halt the fall in real wages. In 1914, the Miners' Federation, the National Union of Railwaymen and the Transport Workers' Federation established the Triple Industrial Alliance, capable of exerting extensive pressure through coordinated strike action. If war between the European nation states had prompted the government to inaugurate a form of state industrial welfare, as explored in Chapter 1, the prospect of 'industrial warfare' between workers and employers provided fertile ground for the development of industrial welfare and industrial psychology.[5] The industrial system was perceived to be in crisis; the factory had become a battleground in which trade unions and employer associations inflicted grievous harm, not only on one another but on the material and emotional well-being of the nation. These anxieties about the explosive state of industrial relations and the possibility of class conflict and ensuing civil strife can only have been exacerbated following the October Revolution of 1917.

The demonstrable failure of the TUC and moderate trade union leadership to rein in their more radical members and recalcitrant shop stewards during the First World War indicated to many that industrial warfare was far from over. Employers and middle-class reformers were terrified by the spectre of an angry and alienated workforce with no loyalty to their employers and no sense of citizenship and duty. Workers also complained that modern industrial practice induced a sense of alienation. 'Vast hives of industry have taken the place of the little factory', complained one trade unionist. 'Men have developed a tendency to become, not living personalities of whom the employer is personally conscious, but mere cogs in the machine.'[6]

To attain greater efficiency and productivity, managers were urged to treat their employees as individuals with a unique psychological disposition. While particular groups within the industrial workforce were explicitly targeted as being susceptible to specific physical or mental health risks, the fundamental principles underlying the incipient industrial welfare and industrial psychology movement in the First World War and interwar years was that all workers should be understood and treated as individuals with a unique psychological make-up, skills and aptitudes if the forces inherent within the workforce were to be harnessed for maximum efficiency and productivity. Industrial relations, one contributor to a 1929 volume on industrial psychology asserted, 'depend essentially on the interests, impulses, sentiments and passions of human beings'.[7]

In practice, the limited resources of this movement focused not on addressing the needs of every individual worker but upon disparate groups perceived to be at the margins of the industrial workforce, linked together by a belief that members of these groups had individualised needs. From the outset, therefore, the quest to address and meet the needs of the individual was inherently in tension with the belief that individual needs could be surmised on the basis of an individual's membership of a designated group within the labour force. Specialised needs could be premised on physiological grounds, notably with reference to a worker's age or gender, or on psychological grounds, with reference to temperament, types of work and working environment. Alternatively, the functions of an individual as a citizen, housewife or future parent outside of the factory could also predicate specialised intervention. This chapter considers how the individual was constructed within industrial welfare and psychology.[8] It will then explore how attempts were made to address the needs of three groups within the workforce: workers whose mental health was perceived to be threatened by the working environment, young workers and women workers.

Constructing the individual in industrial welfare and industrial psychology

In the nineteenth century, Steve Sturdy argued, workers' bodies were perceived as a form of capital which workers invested in employment in exchange for wages, higher wages being received for work more damaging to the body. The demands placed on industrial production during the First World War, however, forced a reinterpretation of the capital of the body.[9] It was no longer acceptable to retain working conditions which undermined physical health or to ignorantly squander the worker's long-term capital by forcing a pace of work oblivious to physiological well-being. The Health of Munition Workers Committee (HMWC) promoted a new physiological conceptualisation of the worker's body and welfare supervisors were employed to ensure that working conditions and hours were arranged to obtain maximum output without damaging workers' bodies. In this process, the specific physiological needs of certain groups of workers – particularly young workers of both sexes and women workers – were subjected to detailed scrutiny. By the 1920s, the needs of the individual worker, overlooked in the development of the factory system, were at the foreground of industrial welfare literature. Between the two World Wars, attention to the physiological needs of individual workers was displaced by a focus on their psychological needs.[10]

Our understanding of the extent to which new psychological ideas permeated and shaped practice within factories has been enhanced by two studies exploring the role of psychology in twentieth-century Britain. Nikolas Rose's monograph explored how psychological thinking was applied to facilitate the government of people in various spheres of life, including the workplace, without transgressing liberal, democratic principles.[11] Rose argued that new psychological ideas regarding the factors that motivated workers led some reformers to attempt to restructure the process of work and the nature of the workplace.[12] Their objectives were twofold and entwined: to humanise work and make it more pleasurable for the worker, and consequently to enhance efficiency and output and so improve employers' profits. Rose's account explored the physiological conceptualisation of the worker during the course of the First World War and the subsequent reinterpretation of the worker as an individual with a unique psychological profile, who would perform best when working in a job fitted to his or her temperament. Attention shifted from the needs of the individual when the human relations approach developed by Elton Mayo began to explore the dynamics of group functioning. Industrial welfarism, with its focus on individual needs, the welfare of young workers and working conditions, consequently began to be displaced in the interwar years by a new model of personnel management which stressed its scientific expertise and neutrality. Following the Second World War, Rose argued, the main unit of analysis became the working group rather than the individual. Efforts were made to democratically manage the workplace, and the interdependency of management and workforce was stressed. This human relations model lost favour in the 1960s, Rose asserted, when the relationship between employee satisfaction and productivity was challenged and the worker was once more conceived of as motivated primarily by pay and promotion rather than social satisfactions.

In Rose's account, the emergence of industrial welfare and personnel management resulted from the application of new psychological concepts to industrial practice; his work offers a different interpretation to that arrived at by historians who have analysed these developments largely in terms of industrial relations and workplace control.[13] Research undertaken for this monograph largely supports Rose's contention that concern with the welfare of the individual worker was gradually displaced in the aftermath of the Second World War by an interest in group relations. Surprisingly, however, Rose gave very little consideration as to how concerns with workers' physical and mental health bisected with the aims and objectives of industrial welfare, aside from a line in his analysis of the 1970s Quality of Working Life movement, where he asserted that 'doctors and others preoccupied with the safety and health of the worker in the workplace and the consequences of work for physical and mental health found a new impetus for their somewhat unfashionable concerns'.[14] While industrial health might well be seen as a peripheral subject by the 1970s, as this account seeks to demonstrate and explain, it is hard to understand why Rose marginalised the health implications of industrial psychology in the earlier period. Moreover, because Rose focused upon developments within the field of psychology rather than on developments within industry more generally, at times his account posits a chronology which is hard to substantiate, overstating some innovations while failing to recognise the longer historical antecedents of other developments. Rose's account exaggerates both the extent of state intervention and the development of the field of industrial welfare work in the years before the First World War, for example.[15] Rose discussed how workers were targeted through new interventions but did not explore how these interventions were gendered to meet the perceived needs of women workers. In one section of his study, Rose argued that in the 1960s there was a growing recognition that workers had a life outside of the factory.[16] Conversely, this chapter will demonstrate just how intertwined an understanding of the home and the workplace had become by the 1920s, reflecting in part the interrelationship between consumption and production explored in Chapter 2,[17] and will argue that industrial welfare supervisors responded to the perceived needs of women workers resulting from the 'problem' of the home and factory. Likewise, Rose asserted that in the 1970s, a progressive politics of working life created a 'new' image of the worker as a unique individual seeking a personal meaning and purpose in their work. Rose argued that the 'old' language of human relations had been radicalised.

> There was an explicit concern with the deleterious social and political consequences of alienation at work brought about by the dehumanising industrial culture... there was a vision of the advance of technology, with its possibilities for the destruction of jobs and the subordination of people to machine on the one hand, but the prospect of the rosy dawn of a post-industrial society liberated from repetitious and uncreative labour on the other.[18]

While this may have been a new approach in human relations, William Morris expressed virtually identical sentiments in 1884 in his indictment of the factory system and his aspirations for a new utopian system of factory production in which mechanisation would liberate the worker from uncreative work.[19] Indeed, it appears that phenomena which Rose identified as uniquely linked to industrial psychology actually stem from a more long-standing discontent with the industrial system. What is perhaps new is the psychological twist given to such long-standing grievances.

This subject receives attention in Mathew Thomson's recent account, *Psychological Subjects*.[20] Thomson explores the broader impact of and interest in psychology and his analysis consequently studies not just the development of industrial psychology but the extent to which workers developed an interest in practical psychology, which offered a holistic approach to enhancing mental and physical health. Both sides on the battlefront of industrial relations increasingly drew upon the language of psychology to bolster the legitimacy of their claims and undermine the demands of their opponents.[21] Class conflict, workers could argue, was a natural psychological response to economic and political injustice. Conversely, as industrial welfare advocates often argued, such conflict should be resolved by brokering a greater understanding between worker and manager. Thomson asserted that a historiography focused on class politics obscures the ethical dimension of psychology in industry and the extent to which reformers at the time were genuinely excited by the potential embodied by psychological ideas to transform the factory system of production and humanise work. Nevertheless, he acknowledged that in practice the remit of industrial psychology became 'disappointingly narrow'.[22]

Mental health in the factory

In the interwar years, a number of writers considered the psychological impact that monotonous and repetitive factory work could have on nominally healthy employees. These investigations were tempered, however, by research which suggested that unemployment was potentially more damaging still to mental health. As the economy slipped into depression in the years following the First World War and the number of unemployed rose, researchers began to explore the psychological impact that unemployment had on people. The influential study undertaken by Marie Jahoda, Paul Lazarsfeld and Hans Zeisel into the effects of unemployment in the Austrian town of Marienthal, in which three-quarters of the resident families depended upon unemployment payments for their livelihood, depicted a vicious cycle in which unemployment bred apathy and despair, undermining the ability of the unemployed to seek work.[23] Similarly, *The Memoirs of the Unemployed*, published in 1934, sought to elucidate the psychological trauma and resulting deterioration in mental and physical health which the unemployed experienced.[24] These studies have been critiqued by Ross McKibbin, who argued that what researchers identified as the 'pathological' characteristics of unemployment were no more than the characteristics of working-class life.[25] Moreover, as McKibbin identified, it was difficult to distinguish the effects of unemployment on health from the effects of the poverty which resulted from unemployment. Families reliant on unemployment benefits would find their budgets for food and housing curtailed, undermining physical health.

For John Gollan, National Secretary of the Young Communist League,[26] high levels of unemployment were simply further indications of an industrial system in crisis. In his 1937 study, Gollan asserted that industrial capitalism 'has established an industrial reserve army – the unemployed; and the existence of this reserve army is essential in order that capital may have a surplus of producers which it can draw upon when needed'.[27] Gollan argued that young people experienced the psychological trauma of unemployment more acutely, citing a recent study on the effects of juvenile unemployment in Wales: 'If it is tragic that a man of fifty should be unemployed, how much more tragic is it that lads of fifteen or eighteen should not be given the opportunities of creative self-expression which would be theirs if they could feel that they were performing some useful function in society.'[28]

If unemployment was so detrimental to health, a handful of psychiatric social workers argued in the interwar and immediate post-war years, perhaps people suffering from mental illness could be helped to recover if they were found work?[29] Although saving government money was also a commonly cited objective, these initiatives were premised on the grounds that employment could contribute to a person's mental health, and their architects stressed the need to find people work which matched their aptitudes and skills, a line adopted by the advocates of vocational guidance. Nevertheless, in practice it was difficult to match the emotional needs of clients with the employment opportunities available in a depressed employment market. The shift away from skilled labour towards repetitive production processes accelerated in the interwar era, provoking concern among industrial health and welfare experts about the psychological impact on healthy workers of repetitive, unskilled work. Even the Assistant Secretary to the TUC, Fred Bramley, argued in the *Journal of Industrial Welfare* in 1921 that 'a good deal of industrial unrest is due as much to mental revolt against the monotony and lack of variety in modern industrial production as it is to evils relating to wages or hours of labour',[30] while a school inspector argued that 'the curse of our industrial system is not drudgery but *meaningless* drudgery; at any moment there are millions of people doing work for their living without the least conception of its significance'.[31] As explored in Chapter 2, industrial welfare advocates argued that it was possible to transform the workplace to help alleviate these problems by making the factory a more beautiful and stimulating environment.

If it was not possible to remove repetitive and uncreative work tasks, companies could provide alternative sources of satisfaction and interest in workers' lives outside their employment. The general adoption of the 48-hour week in many trades for much of the interwar period facilitated workers' opportunities to find fulfilment in their leisure time.[32] Five-day working weeks were also becoming more common and were advocated on health grounds, with greater benefits ascribed to long weekends of leisure to facilitate recuperation from work. This was seen as particularly valuable amid fears of 'speeding up' in working practices.

Another solution to the problem was to identify workers believed to be psychologically suited to repetitive work. During the First World War and in the immediate aftermath, many commentators argued that women were remarkably resilient to the numbing effects of repetitive work.[33] Dorothea Proud claimed that women were able to 'shut themselves out of their work and shut their work out of their lives... curiosity as to their work and their accustomed surroundings seems non-existent... It is as though an unskilled worker brought into the factory only a portion of her consciousness.'[34] Factory Inspector Dr Sybil Horner reported in 1933 that women's adaptability and their ability to balance attention and detachment enabled them to carry out monotonous work on a daily basis without losing their interest in life and risking their mental or physical health.[35]

The Industrial Health Research Board (IHRB) made an attempt to scientifically investigate the impact repetitive work had on an all-female workforce in a study on fatigue and boredom in repetitive work.[36] Noting that their figures were a conservative estimate, the study authors found that over half of the surveyed workforce was moderately or severely bored and that workers were more interested in the prospect of leaving work than in the actual performance of work. Monday was found to be the least-liked day among employees and Saturday most liked. Between 77 and 97 per cent of the workers interviewed stated that they were able to think of other things while working, although for a substantial minority of these respondents their daydreams were far from pleasant distractions from the task at hand. Some workers stated that they spent their working day worrying that they would become unfit for work and be unable to earn money, an anxiety no doubt provoked by the high levels of unemployment, while others wondered how they could 'get out of this hole'. Repetitive work provoked something of an existential crisis on one respondent, who 'dwelt repeatedly on the question as to "why are we in this world at all, since we just go to work and go home to sleep"'.[37] This worker may have found herself in agreement with John Gollan, who linked the rising number of suicides between 1911 and 1933 to the mental damage inflicted by intensification of work, repetitious work processes and the rising levels of unemployment. 'Death is only terrible in proportion to the value we set on life', he argued.[38]

Trade unionists rejected the idea that women were naturally suited to repetitive work. Instead, they expressed concern that such work could damage girls' health. At the 1936 Annual Conference of Representatives of Trade Unions Catering for Women Workers, resolutions on systems of payment for automated processes and provision of more breaks for women and young persons discussed the crushing effect of repetitive work. Miss Horan expressed her concern about conveyor belts, 'which she described as being utterly soul destroying, because the operatives had to keep pace with the machine'.[39] 'I have noted from a close study that girls, very healthy and full of vim, after a few weeks of repetitive work on machines seem to lose all interest in work', one shop steward wrote in response to a health survey carried out by the Amalgamated Engineering Union. 'They become part of the work and appear to be incapable of thinking. They become subject to nerves. Financial worries tie them to this... One thing all have in common is that when they get home they are unable to do anything but sleep.'[40]

If suitability for repetitive work could not be premised on the grounds of gender alone, some hoped that vocational guidance would be able to identify a stratum of workers temperamentally suited to such tasks. In a survey of the literature on the subject of vocational guidance, an Industrial Fatigue Research Board report of 1921 stated that 'a boy or girl found to be naturally unsuitable for one occupation, will almost certainly be found suitable for another. Different occupations clearly demand different aptitudes'.[41] Vocational guidance, the Report claimed, could 'produce a situation in which the worker finds more interest in his work' by 'determining the psycho-physiological characteristics required for a given occupation' and matching individuals with these characteristics to such jobs.[42] In practice, this supposedly scientific method of identifying ideal candidates for repetitive or monotonous work frequently boiled down to the assumption that stupid people were well suited to undertake monotonous tasks. The 1921 Report asserted that physical strength was by no means the only desirable attribute of an employee engaged in heavy muscular work and stressed the value of specific mental characteristics, quoting F. W. Taylor's assertion that the ideal employee should be 'so stupid and so phlegmatic that he more nearly resembles the ox in his mental make-up'.[43] Intelligent workers were ill-equipped to work in modern factories; one author who had studied the attributes desirable for employees of silk mills stated that 'intelligence is not only not required in modern silk mill for most operations but may even be a detriment to steady efficient work'.[44] As the authors of the study on boredom in repetitive work phrased it in 1937, 'the individuals who are likely to find most satisfaction in simple forms of repetition work are those of a relatively low order of intelligence'.[45] This application of industrial psychology indicates why historians have often attacked the narrow, efficiency-driven objectives of scientific management and indeed why it drew criticism at the time from those working more in a welfarist tradition. Throughout 1924, for example, concerns that scientific management would nullify the humanising effects of industrial welfare were frequently expressed in the editorial column of the journal *Industrial Welfare,* and one column criticised the work of C. S. Myers in 1924, claiming that he had asserted that unintelligent people were the best workers on routine tasks.[46]

Myers attempted to defend his work, arguing that while some mechanical work could not be made sufficiently interesting to the more intelligent worker, less intelligent workers would not object and would produce a greater output.[47] Was it possible, however, to identify enough 'stupid' people to undertake boring work? Increasingly both industrial welfare supervisors and personnel managers reluctantly acknowledged that even if vocational guidance was widely applied, not everyone could have a job which they enjoyed and found fulfilling. The authors of the Report on fatigue and boredom in repetitive work concluded that little could be done to transform work to meet the creative and constructive needs of its employees. They found an 'appreciable gap' between the personal desires of workers and their satisfaction in work.[48] If workers were free to choose, 'there is not the least doubt that repetition work would have few adherents', and consequently 'for many individuals, a certain amount of boredom is an inescapable condition of modern factory life'.[49] Instead, the Report suggested that employers should focus on providing distraction outside work and improving the working environment. The 'modern factory', with windows instead of walls, pleasant colour schemes and a spacious interior, was believed to be able to counteract many of the harmful psychological effects of repetitive work.[50] Music in particular was viewed as an effective 'antidote' to boredom and a regular rhythm could encourage workers to maintain an even rate of production. 'The automatic worker is not bored when her mind is full of the dance which she enjoyed or hopes to enjoy', the Report argued, replicating the stereotype of the frivolous factory girl which pervaded the health advice literature throughout the interwar years.[51]

Ultimately, the individual was expected to adapt to the working environment, and the suggestion that the workplace environment might be adapted to meet the psychological needs of the individual, which would ultimately be urged by psychiatric social workers seeking to reintegrate their clients into employment in the 1950s and 1960s, was strikingly absent.[52] The authors claimed to find among their respondents a 'philosophical resignation' to the prevailing conditions among their respondents, a characteristic they believed was accelerated by 'the dulling effects of monotonous work'. 'It is a form of adaptation which prevents conflict and promotes a peaceful frame of mind', the Report insisted.[53]

If it was not possible for all workers to find fulfilment at work, some believed there was a danger that work itself would produce psychological maladjustment and neuroses in previously healthy individuals. Doctors seeking financial support for various schemes to assist and rehabilitate neurotic workers hoped to entice employers and insurance companies to fund their work on economic, not philanthropic grounds. J. R. Rees, Medical Director of the Institute of Medical Psychology at the Tavistock Clinic, argued that psychoneurosis were responsible for one-third of all sickness and cost businesses 10 million weeks of working time each year through absenteeism and inefficiency. He cited the assertions of Dr Lockhart that in some industries as much as 60 per cent of time lost through sickness could be attributed to psychoneurotic illness. Businesses, he argued, should contribute 'largely as a business proposition', claiming that 52 per cent of the cases treated by the Institute were industrial workers.[54]

Neurotic workers were also a subject of interest to the TUC in the 1930s, which investigated the possibility of funding rehabilitative services. In an article and memorandum advocating mental hygiene clinics for industrial workers sent to the TUC, Rosenfield argued that the mental health of workers was being neglected.[55] He suggested that such clinics could investigate, diagnose and treat workers suffering from mental illness, provide psychological rehabilitation to enable workers to regain employment, follow up nervous workers and explore the relationship between neuroses and social conditions such as wage rates and the cost of living. Working conditions, hours of work and methods of remuneration, Rosenfield argued, could cause mental illnesses.

Rosenfield particularly emphasised the plight of workers suffering from neurotic symptoms after accidents at work who were often treated as malingerers and feared that they would be made redundant. Similarly, Dr E. N. Snowden's proposals dealt with the need for psychological rehabilitation for workmen after accidents. Snowden had worked both with cases of shell shock and in the psychotherapy department at St Bart's and he compared the trauma suffered by workers in accidents to the trauma suffered by soldiers in war. He argued that the psychoneuroses were often not recognised as a full disablement and that consequently workers did not receive fair compensation, while the stress of fighting for fair treatment and the worker's belief that he was perceived as a malingerer could worsen a psychoneurotic condition. Snowden proposed that injury cases exhibiting signs of functional disablement be referred to psychologists much earlier and argued that workers should be encouraged to come back into the workplace while they were recovering.[56]

Morgan had asked Snowden to title his paper 'Injury Psychoneurosis (Traumatic Neurasthenia)' as he felt that Snowden's original title, 'Mental Rehabilitation', implied that the memorandum dealt with cases of insanity after accidents. The TUC thus sought to demarcate mental health problems precipitated by accidents, and arguably therefore rooted in physical injury, from the stigma of more severe mental illness, and neurasthenic workers from the insane who required asylum treatment. However, statistics provided by the Amalgamated Engineering Union regarding members who were receiving treatment in mental institutions contained within the file revealed that one in 11 members currently on benefit were either suffering from what was termed nervous trouble, or were in institutions.[57]

While the TUC shied away from acknowledging the extent to which industrial practices produced neuroses, wary of the stigma still attached to mental illness, other writers attempted to arrive at a more exact quantification of the extent of neuroses within industry. A memorandum produced by J. W. Yerrall for the National Association of Trade Union Approved Societies on the subject of nervous disorders suggested that mental disorder was seen as a significant health problem among workers. He claimed that 6 to 7 per cent of employees in large industrial concerns needed treatment for neurosis, while a further 20 per cent had neurotic symptoms that interfered with their happiness and efficiency. Implicating modern industrial practices such as speeding up and excessive overtime in the production of neurosis, Yerrall argued that the introduction of compulsory holidays with pay and more bank holidays would prove beneficial to mental health and consequently prove advantageous to industry. He also advocated improved treatment and facilities for incipient cases in the community such as the provision of convalescent homes.[58]

Russell Fraser found similar levels of neuroses among factory workers in a study undertaken in 1942 for the IHRB.[59] He claimed that 10 per cent of the workers in his survey suffered from what he termed a definite and disabling neurotic illness, while a further 20 per cent were described as suffering from minor forms of neurosis. Neurotic illness, Fraser asserted, accounted for one-quarter to one-third of all illness absence from work. Fraser explored environmental circumstances to try and ascertain what factors might predispose individuals to neurosis. A working week in excess of 75 hours, an inadequate diet and work which the employee found boring or simply disliked were all associated by Fraser with an elevated risk of neuroses.[60] Fraser also found a higher incidence of neuroses among workers assessed to be highly intelligent who were engaged in low-skilled work and on these grounds he explained the disproportionate incidence of neuroses among women designated as highly intelligent, as the majority of women he surveyed were engaged on jobs requiring minimal skills.[61]

Young workers and citizenship

In his first editorial for the *Boys' and Industrial Welfare Journal*, Robert Hyde described the plight of the working boy in industry, whose physical, emotional and educational needs were being overlooked in the pursuit of wartime objectives. From all over the country, Hyde claimed,

> complaints were being made about the working boy; juvenile crime was getting out of hand; jobs were thrown up for trivial causes...Directors, managers, and foremen were being harassed on all sides for greater output...and amid the hustle and rush the boy was in danger of being overlooked, not only as a future workman, but as a future citizen and father.[62]

While the welfare of women munition workers during the First World War has received attention from historians,[63] there has been no comparable investigation as to how boys working within industry were targeted by the State as an industrial group which would benefit from a specialised form of welfare supervision. Speaking at a conference of boy welfare supervisors in 1917, the MP G. Kellaway, secretary to the Ministry of Munitions, claimed that:

> War devastated the manhood of the country, and hardly touched the womanhood of the nation. The toll of war in life, limb and sight made the preservation and development of boyhood a sacred duty of the State...It was their desire, therefore, to liberate the boy's soul, fill his mind, and develop his body to the fullest extent.[64]

These concerns had prompted the Ministry of Munitions to ask Hyde and Seebohm Rowntree to initiate a scheme of boys' welfare work in March 1916.

Harry Hendrick's study of the male youth problem in this era provides one framework through which to analyse these welfare initiatives.[65]

Hendrick explored the concerns expressed by middle-class reformers, educators and medical practitioners within the social, political and economic context of the 1880s to 1920s, arguing that the ongoing labour unrest and industrial disputes between employers and trade unionists and concerns about the place of boy labour within the labour market helped shape the agenda of reformists. Some efforts to improve health and character were focused within schools, such as the provision of the school medical service and physical education.[66] However, it was the fate of boys once they had left school that was of great concern and it was believed that many boys between the ages of 14 and 17 entered so-called 'blind alley' occupations which offered little training, no moral supervision and few prospects, thus damaging their future employment and earning ability. The families of such men, it was believed, would be unable to support themselves and the cycle of poverty would perpetuate itself. Concerns were also expressed about juvenile delinquency, as crime rates rose during the First World War.[67] Anxieties about blind alley work persisted in the interwar years as unemployment attenuated the employment opportunities available to young workers. In his 1937 critique of how industrial capitalism undermined the health of young workers, John Gollan characterised the options open to young workers between the ages of 18 and 20 as blind alley work, monotonous work or unemployment.[68]

From the perspective of many middle-class reformers and commentators, the character of working-class boys had important ramifications for the political stability of the country. Could democratic participation be entrusted to a working class who were only aware of their rights as citizens, not their duties? Hendrick analysed how psychological ideas were used to help create the concept of adolescence and justify interventions such as youth organisations, continuation schools and a youth employment service.

Hendrick's argument is persuasive and an analysis of the welfare schemes established within industry to cater for workers, which were not explored by Hendrick, appears to substantiate his argument to a large degree. Arguably, however, much of what Hendrick identified as the 'male youth problem' could more accurately be described just as the 'youth problem', particularly in the aftermath of the First World War, when the *Boys' and Industrial Welfare Journal* began to cover the needs of girls and women in industry and was correspondingly re-titled the *Journal of Industrial Welfare*. Anxiety about the future social, economic, moral and political roles of young people and efforts to address these concerns through education, exercise, supervision at the workplace and a holistic approach to health was directed at both young men and women working in industry.[69]

One highly contentious issue was the relationship between the wages paid to young workers and their health. Giving evidence to the HMWC, Captain Agnew described one factory where no night work was undertaken and the factory was of a modern design; his account implied that boys benefited more physically and mentally from the conditions in which they worked rather than the wages they received for their labour. 'Their wages were low', Agnew acknowledged, 'but in spite of this fact they all managed before leaving home to have tea, bread and butter, and frequently an egg, though many of them had to leave at 5 am to reach their work at 6 o'clock.' The boys at this factory, Agnew asserted, spent their spare time indulging in boating, swimming, cricket, football, golf caddying and cycling, and were of a 'healthy and intelligent' appearance. Boys at other factories where these working conditions did not prevail, Agnew claimed, were showing definite signs of 'wear and tear' and were 'pale, anaemic, dull and expressionless'.[70]

Other writers believed that wages played a more central role in the health of young workers, enabling them to purchase the healthy diet necessary for sustained work. Factory Inspector Rose Squire related how a large East End restaurant which served food to workers from a nearby restaurant ran out of meat and vegetable dinners one day soon after the outbreak of war, after an unprecedented demand from the younger girl workers who had traditionally opted for the cheaper pudding and gravy or tea, bread and butter. 'The cause was that the wages had that day been raised voluntarily by the employer to the proposed Trade Board rate, and the effect was immediate', Squire noted. 'This fact is a striking answer to those who cling to the theory that an increase in wages is of no substantial value to a girl.'[71]

Agnew's 'healthy' boy workers reportedly started work at 6 in the morning. Other commentators increasingly questioned how young workers were supposed to undertake work, engage in leisure pursuits and still have time to undertake further educational studies if they were obliged to work long hours.[72] This matter was repeatedly stressed in the interwar reports of the Factory Inspectorate.[73] Medical Factory Inspector John Bridge, writing in the 1935 Report, expressed concerns that long working hours, the 'hustle and bustle' of modern working conditions and speed of industry could be detrimental to the health of young workers.[74]

Sport was advocated for young workers of both sexes by welfare advocates, who stressed not only its health-giving properties but its supposed powers to strengthen and discipline character. Through team sports, it was argued, young workers learnt to 'play for the team' rather than personal renown. Cricket and football fields were construed as 'schools of discipline', through which consideration for others and 'a constructive character' could be inculcated and the energies of boyhood, 'always overflowing, and very often, if led to develop of themselves, lamentably misplaced', could be safely channelled and diffused.[75]

For girls, gymnastics was believed to be a particularly effective method to improve conduct and discipline, if not poise: teachers were warned by Dorothea Proud in a book which urged the expansion of industrial welfare throughout industry, that 'a class of factory girls can be the bane of a teachers' existence, if finished grace is the chief object'.[76] Beatrice Webb suggested that hockey was an eminently suitable activity for girls on winter weekends, 'giving concentrated exercise and the best kind of moral discipline, that involved in "Play the game!" – discipline of a kind which comes less readily to factory girls, each working on her own'.[77] Replace 'factory girls' with 'factory boys', and Webb's passage could have been transplanted into a passage on sport for working-class boys.

The risk-taking, undisciplined behaviour believed to be characteristic of young workers was perceived as an attribute which would not only undermine workplace discipline but could endanger young workers' health if they were not adequately supervised. Investigations undertaken by factory inspectors found that young workers of both sexes were 50 per cent more likely to injure themselves when carrying heavy loads than adult women workers.[78] Other reports expressed concern that young workers were needlessly injured because they received insufficient training in the safe usage of machinery.[79] While the accident rate among all industrial workers rose exponentially during the Second World War, the Factory Inspectorate noted that the rate of accidents for male workers was higher among those under the age of 18.[80] Discussing a number of cases which had occurred in lunch breaks, the Inspectorate urged greater supervision and the establishment of lunchtime clubs for young workers which could provide 'counter-attractions to getting into mischief in the meal interval'.[81]

Welfare supervisors and factory inspectors also argued that the educational needs of young workers should be examined and met at the workplace if young workers were to become responsible citizens. Under the provisions of the 1918 Education Act, children under the age of 14 were required to attend school, and would subsequently be required to attend 320 hours of annual continuation schooling until the age of 16. These new requirements spurred the provision of education within the workplace, although some firms had already initiated their own educational provisions in anticipation of the new legislation. In their annual Report of 1918, for example, the Inspectorate described schemes set up by a woollen manufacturer, a confectioner and a cotton firm, all of which incorporated physical exercises such as gymnastics. These schemes provided separate education for girls and boys, offering boys classes in subjects such as English, geography and history, while girls were instructed in housewifery, cookery and sewing. The reports also gave examples of evening classes organised, some of which were oriented more towards leisure activities than the educational schemes.

Women workers, home and workplace

The existing historiography suggests that women workers in the late nineteenth and early twentieth centuries were targeted through legislation because of the perceived threat that industrial work posed to their reproductive capacity.[82] However, as the previous section indicates, the perceived health needs of young men and young women working within industry were strikingly similar. Indeed, when welfare supervisors, factory inspectors and doctors did discuss the specific health needs of women factory workers, the main problem women workers were believed to face was not damage to their reproductive system but difficulty balancing the demands of a full-time job with their housekeeping tasks. This was particularly the case after the cessation of the First World War, a period when militaristic objectives threw anxiety regarding the health of the race into the spotlight. In the interwar years, attention shifted to the relationship between workers' home and work lives and their state of health. In part, this reflected the growing concern that workers generally were unable to find fulfilment through work and would need to compensate for this in their leisure time and home lives. Bored women factory workers interviewed for the IHRB's study of the effects of repetitive work were found overwhelmingly to daydream about home. Many found their thoughts during working hours drifting on to the subject of their evening activities, 'home affairs and everyday matters'.[83] Russell Fraser's investigation into the extent of neurotic illness among factory workers found the main predisposing factor was not the nature of the work or the working environment but a lack of social satisfactions outside of work and, in the case of women, an absence of domestic duties.[84] The campaigns for shorter working hours and holidays with pay also reflected a growing recognition that workers' health and efficiency in the workplace was enhanced by adequate time for leisure activities.

Even during the First World War, concerns regarding the health of women factory workers were divided along two lines. Some expressed anxiety that industrial work undermined women's future role as mothers, an extension of pre-existing social and political concerns surrounding the birth rate and the future of the race which shaped the development of maternal welfare services.[85] Thus one article from *The Times* discussed the need for industrial welfare to protect women as future mothers of the race, although even in this instance emphasis was placed upon the amount of work and inadequate nutrition as the factors which precipitated a breakdown in health. 'It is quite evident that an overworked and underfed woman must sooner or later break down. When she does break down her health may be permanently undermined, and thus a potential mother be lost to the State.' The article warned that '...we are spending womanhood in the factory almost as fast as we are spending manhood in the field'.[86]

Others commentators stated that it was the double burden of maintaining a home and working in a factory, particularly given the long working hours operational during the First World War, that undermined women's health. Industrial work, these writers maintained, could have a detrimental effect upon women simply because many women were unused to industrial work and failed to take enough exercise or eat properly. In her advice book, *Simple Health Talks with Women War Workers*, Sarah MacDonald argued that women were more prone to nerves, anaemia, digestive disorders and headaches, as well as menstrual disorders.[87] Another article from *The Times*, which urged an expansion of welfare provisions including the employment of nurses and welfare supervisors and the establishment of factory canteens, argued that the difficulties women faced lay in the dual nature of the work they undertook within the home and the workplace:

> Woman has the double burden of family life and industry to carry. Unless she is to be cruelly over weighted, the unique and precious service she renders to the State within her own home must be taken into account and safeguarded when determining her position in industry.

Well thought-out welfare provisions which maintained women's health within the workplace and paid due attention to housing conditions and transport arrangements between the home and the workplace could enable women to meet their duties in the home and the factory and maintain their health.[88]

The disruption caused to the home environment during the First and Second World Wars was also believed to undermine the health of women workers and consequently their efficiency. The expansion of factory canteens in the Second World War was urged partly on the grounds that bombing had damaged many workers' home kitchens.[89] MacDonald's book, written during the First World War, focused largely on how conditions outside of the factory could undermine health. She stressed the need to maintain a clean and hygienic home with a supply of fresh air, though she admitted that 'it is with some trepidation, I must confess, that I speak of the problem of hygiene in the home, or what in these times constitutes a home',[90] describing houses which were full to capacity and beds which were in constant use by shift workers. Even for workers living at home, the long working hours during wartime appeared to provoke a sense of alienation among young women. Mass Observation's study of a factory during the Second World War revealed how women began to feel cut off from family life, returning home after long shifts to find they had missed family events, visitors and communal meals. One girl related how the billeting officer had been round when she was working and the family were talking about evacuees. After she expressed her opinion, her sister told her: ' "You leave it to Mum and me," she said; "it doesn't matter to you who we have and who we don't, you're never there." Just as if it wasn't my home as much as hers.'[91] In both World Wars, some attempts were made to support married women factory workers by providing childcare facilities and schemes to enable them to carry out the family shopping. These resources were of a partial nature, however, reflecting ambivalence as to whether married women, especially those with children, should be working in factories.[92]

Industrial welfarists sought to improve the home lives of women workers by extending provisions outside of the factory. Simultaneously, efforts were made to bring elements of the home environment into the factory. Many of the classes instituted for girl workers as part of educational schemes within factories included lessons to teach girls 'homemaking' skills. To further the girls' study, one cotton firm established a cottage set up as a working-man's home, designed primarily 'to afford instruction to girls in housewifery, but it is hoped it may have an educational influence on the girls' tastes and ideas of what a better-class workman's home can be made'.[93] It was claimed that workers, particularly women, would respond to the environment in which they were placed. Throughout the factory, writers urged factory owners and welfare supervisors to consider the impact that décor could have on the morale and behaviour of workers. Pictures, tablecloths, mirrors and flowers were all advocated, and Dorothea Proud even urged her readers to consider arranging the machinery 'with thought to its appearance'.[94]

Anxiety that workers' health and therefore productivity was endangered by their home conditions remained a common theme in the writings of welfare supervisors in the interwar years. A 1922 article from the *Journal of Industrial Welfare* entitled 'The Problem of Home and Factory' asserted that welfare must extend beyond the home. 'The home-visitor who expects to find the most ordinary elements of common sense in the homes of workers will', the author wrote, 'be disillusioned.' The unsatisfactory state of affairs found in workers' homes, the article claimed, was best remedied by targeting young factory girls and providing them with classes in homemaking skills classes after factory hours.[95]

In the interwar years, as concern about factory girls' future roles as mothers receded, writers increasingly identified industrial employment as a factor which could improve as well as undermine health. Medical Factory Inspector Dr Sybil Horner concluded in 1933 that industrial work was largely beneficial to women's health, arguing that wage-earners would benefit from a higher standard of living and pointing to the salutary benefits of discipline and the interests of factory life, both in and out of work hours.[96] Horner suggested that the factory could be a site of health improvement, arguing that 'conditions in factories are, in general, as good and in many cases better than those of the worker's environment…Factory life certainly raises the standard of personal hygiene among girls.' She listed the quality of canteen food, the availability of dental and medical advice and treatment in many factories and the eliminations and reduction of health risks – often accomplished by excluding women from dangerous processes – to bolster her assertion that women's health benefited from industrial work. Indeed, the only significant downside to industrial work which Horner identified was her belief that it aged women rapidly; this appeared to concern Horner not so much because of the possible impact on health than because of its impact on beauty: 'physical attraction is early attained and quickly lost', she wrote.[97]

The Annual Women's Conference of the TUC did not regard the effects of industrial work on women's health with such sanguinity. Speakers at the 1936 Conference voiced their concerns that automated and repetitive processes could have a detrimental impact on the health of girl workers.[98] Anxious that women within industry might experience avoidable health problems because inadequate research had been undertaken, the 1937 Conference passed a resolution calling for an investigation into the health of women in industry. Subsequently, the TUC pursued this enquiry through an analysis of the sickness records held by approved societies and by requesting opinions from doctors contacted through the Industrial Welfare Society and the Industrial Health Education Society. Unsurprisingly, the doctors approached through these societies advocated the establishment of welfare schemes as an effective means to protect the health of women in industry, stressing particularly the importance of good canteen food, adequate seating to prevent varicose veins, flat feet and poor posture and adequate facilities to dry clothes and prevent rheumatism, colds and respiratory diseases.

Doctors Trotter, Platt and Shannon argued that a well-run welfare scheme would help educate women workers in health matters, demonstrating the resilience of the idea that conditions within factories were superior to conditions in many workers' homes and that, consequently, women could learn to improve their homes through the example set within the factory. 'Women particularly need instruction in common sense care of their health and that of their families', wrote Platt. Similarly, Shannon suggested that a works canteen would not only provide a worker with nutritious food but would educate women workers 'in food values, and the various ways of providing nourishing food, which with the knowledge gained, they can supply to their own families'. Shannon argued that anaemia was 'mostly attributable to outside conditions – bad hygiene, unsuitable and ill-chosen food, lack of fresh air, vegetables, etc.' Trotter implicitly alluded to the health benefits of factory work by detailing the deleterious impact unemployment had on women's health. Platt also raised the subject of women's dual roles in the home and factory: women with no conflicting family obligations were 'healthier and happier when in full employment... It is when the family duties conflict with the outside employment... that the health of the worker suffers.'

The summary paper produced by the National Women's Advisory Committee of the TUC in April 1938 did not endorse the suggestion that industrial welfare and educational facilities within factories could teach women workers to take better care of themselves and their homes. Indeed, the study found a higher proportion of men claiming sickness and disablement benefit than women.[99] The principle diseases which women were found to suffer from – rheumatism, colds and influenza, tonsillitis and laryngitis – did not appear to be linked to gender; instead the summary paper noted that women tended to be paid less than men and consequently had less money to spend on necessities and, because of home duties, often had less time for exercise and sleep. A subsequent investigation sought to compare the levels of sickness among women who were paid when away sick and those not paid.

Anxieties regarding the health of women factory workers during the First World War were articulated through a language of biological determinism which stressed the future roles of girl workers as mothers and expressed concern that industrial labour would damage women's health and abilities within the domestic sphere. Throughout the inter-war years, a broad range of participants – factory inspectors, welfare supervisors and trade unionists – continued to express the belief that women's health could be undermined if work and the workplace were not tailored to meet the needs of women. However, these arguments were increasingly premised on the grounds that social and economic factors, rather than any inherent biological characteristics, combined to undermine the health of women workers.[100]

Conclusion: from the worker to the workforce?

One consequence of the investigations undertaken by the HMWC during the First World War to establish the relationship between the working environment, working hours, health and output was the discovery of the individual worker. Whereas it had previously been assumed that labour was an interchangeable commodity or unit, it was now believed that each individual worker brought a unique set of physiological and psychological traits into the workplace. The recognition of the individual needs of workers – arbitrated by the growing professions of industrial welfare supervision, industrial psychology and personnel management – would, it was argued, enable managers to manipulate the working environment to meet individual needs, enhance adjustment and consequently ensure industrial peace and enhanced efficiency.

The perceived needs of particular groups of workers within industry were informed by broader and often quite disparate discourses which articulated concerns about citizenship and social unrest, maternalism and domesticity, unemployment, demoralisation and alienation. The shift from a physiological approach to a psychological approach, which was identified in Chapter 1 as a development which distinguished interventions in the First World War from the initiatives of the Second World War, also shaped efforts to meet the needs of workers. Investigations into the physiological abilities and need of young workers and women within industry were largely displaced by studies of the psychological consequences of monotonous and repetitive work and analyses of how the behaviour and education of women and young workers could imperil their health.

The new disciplines of industrial welfare and industrial psychology brought the plight of the individual within industry to the fore, urging a greater understanding of individual needs and advocating welfare schemes and supervision to ensure that the individual did not feel that they were simply a cog in the machine. It was somewhat ironic therefore that much of the literature strove to create individuals within the same mould rather than genuinely nurturing people's individuality. Young workers were urged to 'play for the team' rather than for themselves; individuals working on monotonous work were expected to adapt to the work.

An exploration of the transformation of medical practice between 1870 and 1950 undertaken by Steve Sturdy and Roger Cooter provides some answers to this paradox.[101] Sturdy and Cooter argued that the desire to enhance national efficiency underpinned a linked process of rationalisation in industrial and medical practice between 1870 and 1950. The pursuit of greater industrial efficiency had spurred on the subdivision of labour within the industrial workplace and had led some to question whether the form and function of the factory should be transformed to enhance productivity; similarly, a desire for greater efficiency within voluntary hospitals led to the introduction of new managers who sought to rationalise the delivery of medical services by a formal and functional division of clinical labour. As industry turned towards methods of mass production, it was increasingly considered inefficient for one worker to individually produce an article. Correspondingly, as the State strove to provide mass health care, doctors were encouraged to discard a tradition of medical practice in which an individual doctor had provided tailored care to individual patients. These objectives were reflected in campaigns to change medical teaching. George Newman wanted to reform the system so that medical students would discard individualised ways of knowing; Newman's vision of a preventive medical service required a new breed of general practitioner, able to recognise incipient cases of illness and assign these cases to particular disease categories.[102] Sturdy and Cooter asserted that medical practice was transformed to serve corporate and national interests; a new managerial culture, keen to enhance efficiency, fostered a subdivision of labour and a new standardised knowledge within medicine which encouraged practitioners to identify categories not individuals.

Considered in this light, endeavours to transform or manipulate the factory and work processes to fulfil the physical and mental lives of workers appear doomed to failure from the outset. What place was there for the individual in systems of medical practice or industrial production – or indeed, within industrial medical services, a subject that the final two chapters will address – which were designed to meet the needs of the masses? Employers were supposed to meet the emotional and psychological needs of workers at the same time that skilled labour and craftsmanship was being displaced by repetitive tasks which alienated the workforce.

The claims made by welfare supervisors that they could individualise industry and promote industrial harmony lost their appeal in the interwar years as the economic situation transformed priorities. Rising unemployment shifted attention from the individual needs of the worker in employment to the damage done to mental and physical health by unemployment; it also weakened the trade union movement and dampened down the threat of industrial strife. While some efforts were made to humanise industry, reflected in the model of the healthy factory, it is unsurprising in this context that scientific management, which promised to enhance productivity, held more appeal to managers than an idealistic model of industrial welfare. Taylorism did not seek to tailor the workplace to the needs of the individual; on the contrary, scientific management sought to assign each worker to a reductive category to facilitate managerial control and efficiency, subsuming their individuality in the pursuit of productionist goals.

If the individuality of workers was stifled within the workplace, one solution was to promote the satisfaction of workers' needs outside the working environment by advocating shorter hours, paid holidays and opportunities for education and recreation. This approach was informed by a new appreciation of the extent to which workers' home and work lives were entwined. These discussions were also inflected by the politics of industrial health discussed in the previous chapter: welfare supervisors, employed by the management, frequently implied that ill-health stemmed from home conditions, while trade unionists emphasised the relationship between wages and health, urged shorter working hours and raised concerns about the impact of monotonous work.

The industrial welfare movement stressed how welfare workers could restore the personal link between employer and employee which had been lost as manufacturing firms increased in size by focusing on the individual needs of workers. However, from the Second World War onwards the attention devoted to the individual within industry was displaced by the growing popularity among personnel managers and industrial psychologists of human relations, which argued that problems within the workplace were rooted in the nature of the relations between people within their environment, to be remedied by enhanced communication methods and an understanding of the informal working culture of the business. 'The unit of observation is the social relationship rather than the individual', proclaimed J. A. C. Brown in his 1954 work, *Social Psychology of Industry*.[103]

These trends were reinforced by the 1944 White Paper on the National Health Service, which stated that the future National Health Service would provide personalised health care, tailored to meet the needs of the individual in his family and home environment, while occupational health services should focus on providing a safe working environment.[104] Consequently, the belief that the health needs of individual workers should be dealt with by a form of industrial health or welfare service began to decline following the introduction of the National Health Service.

5
A National Industrial Health Service?

An industrial health service within the National Health Service

In 1911, shortly before the outbreak of the First World War, the introduction of National Health Insurance established a national statutory system of health care specifically for workers to maintain their working capacity. Though it was to be provided outside the workplace by panel doctors, this legislation set a precedent and ensuing debates on the question of state health care frequently presumed that workers would be the beneficiaries. Subsequently, the development of factory legislation, albeit in a piecemeal fashion, led to a state service within the workplace. This, however, was based on preventing harm, was poorly staffed and aimed to meet minimum standards rather than foster improvements.

An inquiry undertaken by the Amalgamated Engineering Union in 1944 into the health, welfare and safety of engineering workplaces indicated the inadequacy of these arrangements.[1] The report reflected the recent admittance of women to the union in 1943 following the influx of women into engineering during the Second World War, dealing with facilities for women workers.[2] Undertaken by shop stewards, the Report covered a total of 1,253 workplaces managed by 988 separate firms and employing a total of 1,300,000 workers. The Report highlighted the wide divergence of standards among premises, indicating that many failed to comply with the requirements of the Factories Acts. One shop steward reported that 'the ventilation in the sub-assembly shop is too bad for description. With the smell of rotting, rat-infested wood floors the low roofed, badly overcrowded shop is worse than the Black Hole of Calcutta.'[3] Inadequate washing and sanitary facilities were particular areas of complaint. Explaining the decision to establish the inquiry, an article in the *Amalgamated Engineering Union Journal* expressed the hope that the findings would 'rally ever-widening support for the setting up of a National Industrial Health Service'.[4]

The prospect of a national industrial health service incorporated within a new national health service appeared imminent during the Second World War, when the appointment of industrial medical officers under the 1940 Factories (Medical and Welfare Services) Order promoted a model of industrial medicine characterised by medical supervision of the worker in the workplace. This model would continue to influence thinking among medical organisations and trade unions in the aftermath of the Second World War, though the latter remained concerned with another precedent established during the war: the government's decision that the employment and remuneration of industrial medical officers should be left in the hands of managers rather than the State. This had repercussions on the shape of the service, which consequently focused on the maintenance of industrial production.

A memorandum on medical supervision within factories issued by the Ministry of Labour and National Service in 1946 illustrated how these developments affected the role of the doctor working within industry.[5] This memorandum listed a set of functions which would come to typify many organisations' views of what the role of works doctor or industrial medical officer entailed. It stated that the doctor should be responsible for the organisation of first aid services for sickness and injury occurring within the workplace and should examine new employees and those returning to work after sickness or injury. He or she should also advise the management on issues of general hygiene within the factory and provide health education to the workers.

While the memorandum claimed that an industrial medical service could benefit both employers and employees, it stressed that the works medical officer should be a part of the management. The memoranda also noted that 'the factory may well form the basis of education in hygiene to be translated in the home', a sentiment distinctly reminiscent of the interwar argument that workers' health was jeopardised by their unhygienic living conditions, born of ignorance, and that a modern healthy factory could actually be a site to improve workers' health. Another feature which would become characteristic of the prescriptive role of the industrial medical officer following the inauguration of the National Health Service (hereafter NHS) was also present within the document; the works doctor was urged not to give continued treatment at the works without securing the agreement of the patient's panel practitioner.

During the war, medical organisations published reports and memoranda which outlined their vision of what a future industrial health service should look like. Comparing three such reports: the first produced by the British Medical Association's (hereafter BMA) Committee on Industrial Health in Factories published in 1941,[6] the second a memorandum produced by the Association of Industrial Medical Officers in July 1943 on 'The Place of an Industrial Medical Service in a Post-War Comprehensive NHS'[7] and finally the Association of Scientific Workers' 'Memorandum on the Industrial Medical Services', printed January 1944,[8] it is evident that on most main issues, medical organisations were largely in agreement.

All three of these organisations, for example, emphasised that the fundamental characteristic of a future industrial medical service should be medical supervision of workers, undertaken by the industrial medical officer. A medical practitioner at the place of work should supervise the working environment as well as the personal health of workers. He or she would provide treatment and rehabilitation services to workers incapacitated at work and follow-up treatment of disease and injury. The Association of Industrial Medical Officers, a body established in 1935 to represent the needs of doctors working full-time in industry, unsurprisingly stressed the specialist nature of industrial medicine. In his history of the Association, J. T. Carter argued that the stress placed upon the acquisition of clinical skills was seen as a means to promote recognition of the industrial medical officer as a specialist with an unconstrained field of practice. Acting on concerns that this approach would lead industrial medical officers to encroach on the terrain of general practitioners, the BMA drew up ethical guidelines for industrial medical officers which established a delimited field.[9] The Association of Industrial Medical Officers stated its belief that special training and qualifications for medical personnel working within industry should be required and that nationally organised research facilities should be established alongside machinery to disseminate research findings. The medical organisations also agreed that medical supervision was important within all factories, irrespective of their size. The Association of Scientific Workers raised as a particular area of concern the lack of provision for smaller factories: 99 per cent of factories employed fewer than 500 people, it reported, but most doctors were employed in the 1 per cent of factories which employed more than 500.

All these organisations stressed the need for an industrial health service. Their reports pointed out that the current system only provided medical supervision at the workplace for 25 per cent of industrial workers and stated the belief that any comprehensive national health service would have to provide for the supervision and medical control of all occupational and environmental factors affecting the health of people at their place of work. All agreed that an industrial medical service should be a statutory requirement. There was some dissension however as to whether industrial medical officers should continue to be selected by employers; the Association of Industrial Medical Officers argued for a continuance of this system but suggested that doctors should in future be responsible to an authority administered by the future NHS. Conversely, the Association of Scientific Workers believed that employers were unqualified to judge medical qualifications and suitability. They urged the establishment of a state service on the grounds that continuity of medical services currently depended upon the financial position of the employer rather than the need for the service, referring to the fear 'often expressed by workers, that the medical departments in their own factories will be discontinued after the War'.[10]

The TUC and its affiliated unions had persistently attempted to secure healthier working conditions for their members in interwar Britain, when the unfavourable economic and political climate ran counter to such a policy. It believed that the wartime expansion of medical services within factories and the government's commitment to establish a national health service presented an unparalleled opportunity to secure a national industrial health service. In 1944, the TUC Congress passed a resolution which called upon the government to provide an industrial medical service which would cover all sections of industry and commerce within the ambit of the projected NHS, in which medical personnel would be accountable only to the service and the State.[11]

However, industrial health services were excluded from the NHS, for reasons given in the 1944 White Paper, *A National Health Service*.[12] In a section entitled 'Some misconceptions about the meaning of "comprehensive"', the Report argued that while a comprehensive service should encompass all forms of personal health care, it did not have to incorporate all activities, whether run by the government or private enterprise, which employed medical personal or had a bearing on health. Explicitly referring to the employment of industrial medical officers and the work of medical factory inspectors, the Report acknowledged that medical personnel played a role in ensuring a 'proper' hygienic working environment, allocating work to match the capacity of individual workers and generally protecting the worker's welfare. However, such medical interventions were described by the Report as 'part of the complex machinery of industrial organisation and welfare, and it belongs to that sphere more than to the sphere of the personal doctor and the care of personal health – which centres on the individual and his family and his home'.[13]

The Report introduced a strict divide, asserting that the future NHS would provide personal health treatment and medical advice while industrial medical services ensured the safety of the working environment, and consequently belonged to the scope of industrial organisation rather than personal health care. This divide was crucial in determining the future fate of industrial medical services; it was also a position which the Ministry of Health would later partially retract.[14] If the reports of the Health of Munition Workers Committee (HMWC) from the First World War are revisited, it becomes clear how much this new model diverged from past ideals. The HMWC had advocated a holistic approach to workers' health, subscribing to a model of medicine which was concerned not only with the diagnosis and treatment of disease but with prevention. It had emphasised the psychological dimensions of preventive medicine, suggesting that medicine had to consider the whole person in all spheres of life. The White Paper on the future NHS jettisoned these ideas.

The same passage within the 1944 White Paper dealt a further blow to aspirations for a state-run national industrial health service. Specialised services such as those provided within factories, the Report explained, 'should not impair the unity of personal health service' and consequently, 'where there arises – perhaps first detected in workplace or factory – a question of personal medical treatment or consultation...this should be regarded as a matter for the personal health service'.[15] This section implied that the development of industrial health services could impede the working of the future NHS and sought to limit the scope of industrial medical services, a setback to those industrial medical officers who believed that the workplace was an appropriate site to provide a personalised medical health service. Although this passage might suggest that industrial health care arrangements – 'matters of industrial organisation' – belonged more rightly within the remit of the Ministry of Labour, the Ministry of Health would later expand upon this argument in an attempt to curtail and wrest control over the industrial health services.

Approaching this White Paper from the perspective of industrial health, one might wonder why the TUC did not object more strongly to the exclusion of industrial medical provisions from a national health service. However, while the TUC expressed its disappointment with the exclusion of industrial health services, it accepted the Report as a compromise solution, unwilling to risk jeopardising the prospect of a state national health service in its campaign to secure adequate protection within the workplace.[16] Thus, speaking in the Commons, the TUC's Medical Adviser Dr Morgan welcomed the arrival of the long-awaited White Paper with a mixture of relief and disappointment. 'There has been a long period over it and I must say that I expected ... a monstrosity', he stated. 'At last the infant has been produced; it is not a very healthy infant; it is rather emaciated, emasculated ... However, if only half of what is in the White Paper comes to fruition we will have made a real land-mark in the health services of the country.'[17] Morgan objected that the scheme proposed in the White Paper was not so much a comprehensive national health service but a medical service, complaining that public health and social medicine had been left out. Much of his criticism was reserved for the exclusion of industrial health and he stressed the defects in medical training and what he saw as an inherently flawed attempt to separate preventive from curative medicine and industrial medicine from general health.[18]

In response to TUC pressure, Bevan did explore the option of transferring responsibility for the industrial health services to local authorities in November 1945, only to drop these proposals a month later.[19] Indeed, Bevan was not opposed in principle to integrating industrial health services into the National Health Service. A set of notes compiled by the Ministry of Health in February 1946 in response to the BMA Council's Report on the NHS Bill largely concurred with the BMA's view that the exclusion of industrial medicine was unjustifiable. 'Quite right that it ought to be tackled. No disagreement. Simply that cannot do everything at once. This Bill deals will all *treatment* services. Will tackle industrial medical services (with Minister of Labour) later.'[20]

The future of the industrial health services outside the NHS: the industrial health advisory committee

In November 1944, the Industrial Health Advisory Committee (IHAC) met to discuss the future position of industrial medical services in relation to other medical services. This Committee had been established in 1943 by Ernest Bevin, the Minister of Labour and National Service, after a report issued by the Select Committee on National Expenditure argued that not enough was being done to safeguard the health of workers in wartime,[21] and contained representatives from the TUC, BMA and employer organisations. The meeting illustrated some fundamental differences of opinion which would come to characterise future discussion on the subject.

Dr Goldblatt was a founder member of the Association of Industrial Medical Officers and worked as an industrial medical officer at Imperial Chemical Industries, a firm with a long history of providing medical services at the workplace.[22] Consequently his paper, which largely reiterated the approach taken to industrial medicine by the 1944 White Paper, represented the interests of industry. Reflecting the movement away from personalised health care provisions, explored in the previous chapter, he claimed that as an industrial medical service was concerned with 'not only the health of the employee as an individual but with the employed community', it should be an integral part of industry. An industrial medical service, he asserted, was not detached from but intimately interested in the success of industry.

Charles Hill, secretary of the BMA, criticised the proposed exclusion of industrial medical services from the future NHS. Hill attacked the assumption within the White Paper that the main function of industrial medicine was to regulate environmental conditions: 'modern industrial medicine', he claimed, 'includes an important element of personal supervision and should not be regarded as dealing solely with ventilation etc.'[23] Moreover, Hill argued, it was impossible to separate the factory environment from the general environment, and the health needs of the citizen from the worker. In his view, industrial medical services were incomplete, failing to cover smaller factories, and only a state service, interwoven as an integral part of the future NHS and managed by the Ministry of Health, would ensure development and independence.

J. L. Smyth, the Social Insurance Secretary for the TUC, advocated a comprehensive national medical service for both the prevention and cure of sickness. He urged the incorporation of industrial medicine into the medical curriculum, as the TUC had recommended for a number of years, and the establishment of postgraduate diplomas in industrial medicine. The TUC wanted to see a specialist industrial medical service which was interwoven with the NHS and Smyth suggested that this could be achieved if the Ministry of Health assumed responsibility for industrial health services as part of the NHS, but held the power to delegate the matter to the Ministry of Labour.[24]

A TUC deputation presented these concerns to Ernest Bevin, the erstwhile trade unionist who had overseen the expansion of industrial medical services during the Second World War in his role as Minister of Labour and National Service. Meeting in May 1945 to discuss two Congress resolutions – on the industrial medical service and the revision of the Factories Act – the TUC found that Bevin's time in ministerial office had changed his views. Bevin explained that 'he used to have the same view as the TUC that the doctor in the factory should be an employee of the State', but had subsequently come to the conclusion that 'to relieve the employer of his responsibility for the protection of the health of his employees whilst they were under his control, would be a mistake'. In his view, an industrial medical service was the responsibility of management and ergo should be paid for by management as had occurred during the Second World War, a perspective which he insisted 'had now been accepted by practically everyone in the country', although this apparent broad consensus did not include the TUC.[25]

The IHAC meeting on the future of industrial medical services held in November 1944 had demonstrated a clear lack of consensus as to the function, funding, scope and responsibilities of such a service. Subsequently, the Committee established a sub-committee to consider the development of industrial medical services in factories. This sub-committee distributed questionnaires to the BMA, the British Employers' Confederation (hereafter BEC), the Association of Certifying Factory Surgeons, the Royal College of Physicians and the TUC. The collated responses pointed to divergent views and indicated that it would be difficult to achieve consensus between different parties.[26]

For example, while the TUC felt general medical supervision should be made compulsory in all factories, irrespective of their size or function, the BEC believed it should only be extended on a voluntary basis. The three medical organisations adopted a middle ground, arguing that compulsory medical supervision should be rolled out in stages.

There was little agreement on the issue of how industrial medical officers should be appointed and what medical supervision should entail. The TUC believed medical supervision should be restricted to curative and preventive action to prevent ill-health from occupational causes. It envisaged that the industrial medical officer would oversee first aid arrangements, undertake the examination of young workers and those engaged in special processes and provide care and supervision to safeguard the health of adolescents and women workers; this latter point indicating that the TUC still believed that certain groups of workers had personalised health needs which were best met by a specially tailored service. It should also be the role of the industrial medical officer to supervise occupational hazards and provide health education, the TUC argued, although it envisaged that the enforcement of legal standards of lighting and ventilation should remain the responsibility of the Factory Inspectorate. To ensure impartiality, the TUC felt a government advisory body should appoint doctors to a panel from which selections would be made, with both management and workers having a say in the process. It wanted to see training in industrial medicine extended and felt it desirable that doctors employed as industrial medical officers worked full-time in the field.

The BEC stressed that the selection and appointment of doctors was a matter solely for management. It assumed that the industrial medical officer would supervise general conditions and first aid arrangements but would not carry out systematic examinations of individual workers or provide them with medical or curative treatment; this reflected the shift from the health needs of individual workers to interest in the health of the workplace environment, explored in the previous chapter. Reflecting the interests of its members, the Association of Certifying Factory Surgeons argued that the Factory Acts should be extended to place extra duties on examining surgeons, visualising an industrial medical service which would therefore be staffed largely by general practitioners working on a part-time basis. Envisaging that some of the industrial medical officer's time would be devoted to overseeing the general working conditions, including the works canteen, most of the functions listed by the Association fell within the field of personalised health care: the organisation of rehabilitation, nursing and first aid services, follow-up examinations of young persons to ensure that they were suited to their environment and consultations for workers.

Both the BMA and the Royal College of Physicians urged an extension of specialist training so that full-time industrial medical officers with expert knowledge could work in the service alongside part-time doctors. Both outlined a set of functions for the industrial medical officer which encompassed overseeing the safety of the working environment and the health needs of the workers. The BMA envisaged a fairly broad-ranging remit for industrial medical officers, much of which was targeted at protecting the health needs of individual workers. The industrial medical officer, the BMA argued, should examine applicants and advise managers on selection. All young workers should be continuously observed by the industrial medical officer, who could recommend the provision of free meals or milk if it was felt necessary. Similarly, the BMA felt that the industrial medical officer should examine workers returning after a period of sickness or injury, and examine workers exposed to special hazards, as well as providing treatment for medical emergencies occurring at the workplace. The BMA believed that the health and safety of the working environment was a responsibility of the industrial medial officer, who should advise management on factory legislation, hygiene and potential hazards posed by the working environment, materials or processes. The industrial medical officer should also, in the BMA's view, disseminate health information to works councils and welfare departments, encouraging supervisors to report signs of ill-health in workers.

Likewise, there was no agreement as to which ministry should be responsible for an industrial medical service. The BMA argued that a special department of occupational health should be established within the Ministry of Health which would run the service. The TUC and the Royal College of Physicians both stressed that an industrial medical service should be an integral part of the NHS, administered by the Ministry of Health, which should have the power to delegate the running of the service to the appropriate government department. The BEC and the Association of Certifying Factory Surgeons both stated that an industrial medical service was rightly a matter for the Factory Department of the Ministry of Labour.

A chicken and egg conundrum: staffing an industrial health service

In June 1948 the TUC arranged for a deputation to meet with George Isaacs, the Ministry of Labour and National Service, to discuss its concerns with the lack of visible progress made towards the establishment of a national industrial health service.[27] The sub-committee appointed by the IHAC to advise and report on the development of an industrial medical service in factories had met only twice, the last time over a year previously when it had issued questionnaires to interest groups. 'Assurances have been given on a number of occasions that it is the government's intention to develop such a service', a TUC memorandum noted, 'but there is little evidence of constructive action being taken.'[28]

In the absence of a lead from the government, other developments were taking place in the field of industrial health. Outlining industrial health work being undertaken by voluntary organisations, including the establishment of local advisory committees on industrial health and a scheme to provide medical services on a grouped basis to small companies in Slough, the TUC memorandum noted that 'without in any way commenting on the advisability of these developments, which do at any rate show the need for action, we consider that they make it more than ever necessary that the government should announce its intentions'.[29] The TUC viewed voluntary initiatives as a threat, which might stem the flow of government involvement. It believed that only a state-run scheme could offer comprehensive coverage. The NHS had come into operation in July 1948 and it must have been particularly galling to witness the establishment of a national service which excluded industrial health in the knowledge that no apparent steps were being taken to establish a national industrial health service. As the *Financial Times* commented in that month:

> It was three years ago that the Royal College of Physicians recommended that industrial medicine should be brought within the ambit of the new health service. If this had been done it would have been possible to group sufficient of them [i.e. factories] to have justified either the whole or part-time employment of doctors ... Unfortunately, either because of interdepartmental jealousies, or for some reason known only to the powers that be, this sensible suggestion has been ignored.[30]

At the meeting in July, Isaacs disputed the TUC's accusation that the government had been idle but claimed that it was impossible to introduce legislation until a more substantial foundation had been built up and the acute shortage of doctors with industrial qualifications had been remedied. The TUC felt that this was the wrong way to approach the problem: in its view, unless the government issued a statement announcing its intention to establish a national industrial medical service, doctors would not be persuaded to gain industrial qualifications. Dr Morgan, the TUC's Medical Adviser, wanted to see teaching on industrial health take place in all 42 medical schools in Britain, and the implementation of a national scheme into which all factories would ultimately be brought. Morgan emphasised the danger of waiting for a "quality" service staffed by expert personnel to evolve while workers were running risks from the development of new chemicals and processes.[31]

Subsequently, Isaacs contacted the prime minister and the Minister of Health, requesting their views on a statement of government policy he proposed to send to the TUC. The TUC, he explained, 'have consistently advocated the establishment by the State of a comprehensive medical service for industry', and were 'disturbed because there did not seem to have been much government activity in the direction they desired'.[32] The response Isaacs dispatched to the TUC sketched out the developments that had taken place since the cessation of the Second World War, including the establishment of four diplomas of industrial health and university departments of industrial health at Manchester, Glasgow and Durham. Isaacs also outlined the development of grouped industrial health schemes. On this basis, he asserted that great progress had been made and that the government had worked to a plan. Privately however, Ministry of Labour staff expressed the view that, with regard to the development of industrial medical and nursing services, 'this Ministry may find itself in an embarrassing position if it does not soon take more active steps than it has done lately to promote an expansion of those services'.[33]

Discussion not action

The evidence presented before the Industrial Health Services Committee (IHSC), chaired by Judge Edgar Dale, is explored in the following chapter. Given its terms of reference, the Report of the Committee, published in 1951, was very positive about the role of the industrial medical services.[34] It concluded that industrial health services were very important to industry, complementary to the NHS and should be encouraged to expand with due regard to the demands of other health services for medical manpower. The Report also acknowledged that more research into the needs of industry should be carried out, and discussed the potential of collective schemes serving numerous organisations within a particular locality. Eventually, the Committee argued, there should be comprehensive provision for occupational health covering industrial and non-industrial occupations, a subject that had been under consideration by the Gowers Committee, which had reported in 1949.[35]

These findings contradicted the evidence given to the IHSC by the Ministry of Health, prompting Aneurin Bevan to write to Attlee requesting that both publication of the Report and the lifting of the ban imposed on further developments of industrial health services be deferred. The right course of action, in his view, would be to appoint a working party to study the Report and advise on the 'best ways and means of giving effect to its various recommendations'. The constitution, powers and terms of reference of the two committees envisaged by the Report would, he claimed, 'require more detailed study'.[36] George Isaacs flatly rejected Bevan's attempt to procrastinate: given the unequivocal nature of the IHSC's recommendations, attempts to postpone publication or retain the ban on the development of industrial health services would leave the government 'open to severe criticism'.[37] When Attlee sided with Isaacs, Bevan wrote again expressing his disappointment with the Committee's findings. 'It is not merely that I think they have come to certain somewhat ill-founded conclusions', he wrote, 'but also that they seem to me to have made such a superficial examination of the position, without ever really getting down to the fundamental problems that arise':[38] this assessment of the Dale Report would subsequently be reiterated in virtually every Ministry of Health paper on the subject of occupational health services, prompting Mary Smieton at the Ministry of Labour to scathingly comment, 'no doubt the Dale Committee appeared to the Ministry of Health superficial because it came down on the wrong side from their point of view'.[39]

The Lancet gave the Dale Report a lukewarm reception, expressing its disbelief in a leading article that it should be felt 'necessary to apologise for the existence of the services, which, however inadequately, look after the health of our large industrial population'. It suggested that if effective coordinating machinery was not established to develop industrial health services, 'the Dale report will go down in the history of industrial medicine as another lost opportunity'.[40] This proved to be a rather prescient prediction. Although Attlee accepted the recommendations of the IHSC, his government fell from power before the standing advisory committee proposed in its Report had been established: the IHSC had envisaged that the advisory committee, which would include representatives from the Ministries of Health, Labour, Fuel and Power, employers, workers, doctors and nurses, would coordinate and develop occupational health services. In October 1951, following a narrow Conservative victory, Winston Churchill appointed Walter Monckton to what he described as 'the worst job in Cabinet': Minister of Labour. Faced with the challenge of minimising hostility from the trade union movement following an election in which the Labour Party had secured more votes than the Conservatives, Monckton allegedly replied, 'I take it you do not expect me to have a political future.'[41] Monckton argued that the appointment of the standing advisory committee was not urgent as the IHSC had not unearthed any evidence of overlap between the NHS and industrial health services. Instead, the Ministry of Labour circulated a paper to the National Joint Advisory Council suggesting that this body establish an Industrial Health Advisory Sub-Committee.[42] This paper discussed the possibility of grouping small factories to provide an industrial health service, before pouring cold water on the idea. It queried whether the managers of small enterprises would respond enthusiastically to an invitation to collaborate in a grouped scheme at their own expense, especially in cases where there was no perceived abnormal risk to health or 'the workpeople's needs for a medical service are adequately met by the NHS'.[43] The paper also queried the advisability of establishing more extensive medical supervision in small factories when other workplaces lacked any medical supervision.

'We cannot treat the Dale Committee's recommendations as though there had never been a committee', acknowledged Guildhaume Myrddin-Evans, head of the international division of the Ministry of Labour. 'It seems clear that the BMA will not be content to let the whole thing sleep – nor will the TUC; and if we do nothing we can expect renewed pressure to transfer our responsibility to the Ministry of Health.'[44] Myrddin-Evans' concerns were well-grounded: A memorandum drafted by the TUC in October 1952 expressed frustration, stating that unless a government department could be convinced of the importance of developing industrial health and was prepared to give a vigorous lead, nothing valuable was likely to come from the duplication of committees.[45]

The TUC was unhappy that the recommendations of the IHSC were being discussed by the National Joint Advisory Council, a tripartite body established in 1939 to negotiate matters pertaining to industrial relations. The Ministry of Labour's decision to delegate the question of industrial health services to this body indicated that it perceived industrial health services to be an industrial relations matter; conversely, the TUC viewed industrial health services primarily as a health matter, albeit one which required specialised knowledge. In July 1952, Tewson wrote to Walter Monckton to enquire what the Minister intended to do with the IHAC, which had last met in 1950. Monckton argued that the work of the IHSC had rendered the former body obsolete, but trade unions remained unhappy with the approach adopted.[46] 'The importance of developing industrial health fully justifies a strong and vigorous committee with special responsibilities for the job', a set of TUC notes commented. The notes expressed dissatisfaction that such a 'vitally important matter' had been passed to the National Joint Advisory Committee, which had many other commitments and viewed the proposal as in line with the Factory Department's 'failure to give any real lead in developing industrial health'. 'The real issue is the apathy of the Factory Department in developing industrial health', the TUC notes stated, 'and its failure to move with the times.'[47]

In November 1952, Monckton wrote to Vincent Tewson to express his concern that a TUC representative at a recent National Joint Advisory Council meeting had suggested that the Ministry of Health or Ministry of National Insurance might be a more appropriate department to develop an industrial health service that the Ministry of Labour.[48] Acknowledging that the TUC's ultimate objective might be the establishment of an occupational health service, Monckton asserted that the current proposals related only to the development of industrial health services. In his view this involved the elimination of unhealthy working conditions and fatigue and the employment of industrial medical officers to undertake examinations of young persons and of workers on certain unhealthy processes, to oversee first aid and preventive treatment and to investigate the causes of disease within the workplace. Therefore, Monckton argued, the industrial health service 'is not primarily a personal health service to individuals like the NHS, but is a general preventative service connected with work and work places and is dependent on the carrying out of certain functions, largely non-medical, *at the place of work*'. He added: 'I should hardly think that this view would be challenged.' Should an industrial health service eventually be extended to encompass workers in non-industrial occupations, Monckton argued that it would be governed by the same principles. In his view, an industrial medical service, entailing medical supervision, was simply one aspect of a broader industrial health service, which would be staffed by engineers, chemists and welfare officers. 'So far as the industrial health service is concerned', Monckton wrote, 'the problems are industrial ones and can be dealt with effectively only by an industrial department.'

Through this letter, Monckton hoped to gauge the TUC's expectations. Ministry of Labour comments on an earlier draft of this letter acknowledged that it sidestepped the main point raised by the TUC regarding the possibility of a future comprehensive occupational health scheme which would embrace all workplaces: 'at present there is no such thing as an Industrial Health (or Medical) Service, or even a Factory Health (or Medical) Service'. Contemplating the obstacles lying in the path of a projected comprehensive occupational health service, the notes concluded 'in brief, the wicket is sticky and is not going to improve and I rather feel that the best policy is to let the TUC bowl first.'[49]

The TUC subsequently arranged to meet Monckton in 1953 to express its discontent, armed with a memorandum which noted 'the need for a practical programme is urgent. The Government's apathy and neglect during the post-war period is open to serious criticism.'[50] The deputation aimed to discover whether Monckton intended to take responsibility for the development of an occupational health service or an industrial health service. The existing factory medical service provided cover to only 7 million workpeople out of a total working population of 22 million and the TUC felt that it would be impossible for the Factory Department as it was presently constituted to undertake the provision of a comprehensive occupational health service. The notes also expressed exasperation with the internecine feuding between government departments, recounting the 'general belief that the Dale Committee was seriously handicapped by this inter-departments conflict of views' and noting that a recent leading article in *The Times* had stated that 'it is not right that the present state of inactivity should continue because of the failure to come to grips with administrative difficulties and departmental jealousies'. The TUC was concerned that the Factory Department lacked adequate resources, was understaffed and had failed to give an effective lead in the development of occupational health during the post-war years.[51] Within the Ministry of Labour, one staff member complained about 'the ignorant cry that nothing has been or is being done about industrial health'. In his view, the campaign for a 'regimen of doctors' had been orchestrated by the BMA 'whose representatives know little of industrial health and all of whom have a professional interest in the ascendency of doctors'; the TUC, 'whose vague demands appear to imply that they want a preferential National Health Service for employed persons at their place of work' and the Ministry of Health, 'who for their own purposes seek to control all medical services'.[52]

The unrelenting pressure placed by the TUC upon the Ministry of Labour was beginning to have the desired effect. In October 1953, Ministry of Labour personnel began drafting memoranda for a prospective occupational health scheme.[53] Acknowledging the 'interdepartmental jealousy' such proposals would engender, and seeking to avoid 'a pitched battle', Mary Smieton consciously chose to focus on the industrial side of occupational health while de-emphasising its medical aspects.[54] 'From discussions with the TUC', she noted, 'there is no doubt as to the strength of their feeling that the health of the worker in the place of his employment has not received the attention it deserves, and as to their intention to press this issue as one on which the government has failed.'[55] Monckton enumerated similar reasons in December 1953 when soliciting the support of Frederick Marquis, Chancellor of the Duchy of Lancaster. Expressing his conviction that an occupational health service would be a political success, Monckton warned, 'If we do not do this, and without too much delay, we shall be attacked by the

TUC and I am quite sure that our opponents, if they should get into power, will steal our thunder.'[56] Vincent Tewson maintained pressure on the government by dispatching a letter to Monckton enquiring as to the progress of proposals for occupational health services, using the opportunity of a deputation to the Prime Minister the following day to restate the urgent need for immediate steps to advance occupational health facilities.[57] In meetings with the Ministry of Labour in January and March 1954, the BMA restated its desire for an occupational health service, expressing concern regarding the lack of action.[58]

In 1954, Monckton presented a memorandum outlining a possible national occupational health service to a meeting of the Cabinet. Citing pressure from the TUC and the BMA for a coordinated occupational health scheme, Monckton advocated the establishment of an occupational health advisory council to advise ministers on all aspects of occupational health services and their development, and an inter-departmental committee to coordinate government policy on occupational health, to review the progress of legislation and to arrange regional surveys and experiments. He envisaged that these committees would focus in the first instance on carrying out consultation regarding the recommendations of the Gowers Report, intensifying environmental measures to secure improvement in health in workplaces and stimulating the voluntary provision of occupational health services, as the Dale Report had recommended. Noting that the Labour Party had displayed an interest in inaugurating an occupational health service, Monckton stressed the political advantages which could be reaped by implementing an occupational health service shortly before an election and thus placating the TUC and the BMA.

A counter memorandum presented by Iain MacLeod, the Minister of Health, urged yet another review of the content and provision of occupational health services, lambasting the 'weak and timid' conclusions of the 'superficial' Dale Committee. In discussion, several ministers queried whether it was an opportune time to inaugurate such a service, given the government's efforts to curb expenditure on existing social services. Winston Churchill asked whether it was wise to raise hopes if 'we have no real intention or hope of making progress soon'. Deferring a final decision until the proposals could be considered alongside memoranda prepared in response to the recommendations of the Gowers Report, Home Secretary David Maxwell Fyfe ruminated that 'the time seemed ripe to consider what should be the main theme of Government action during what might prove to be their last year of office, and the Cabinet would wish to consider whether this should be confined to retrenchment or whether this include any measures of social progress'. 'If we decide to do nothing', he warned, 'the Minister of Labour and I will have to handle the trade unions very carefully.'[59] A Ministry of Labour note on the proposals drafted around the same time asserted that:

> The government must now either decide on the promotion of a National Occupational Health Scheme, or, as an inescapable alternative, take the blame for doing nothing. Both the BMA and the TUC have given lengthy consideration to this question and are pressing various proposals on us. Both are concerned at the delay and lack of action on the part of the government. It would be unrealistic to expect them to continue to refrain from public and political criticism.[60]

The Cabinet reconsidered the proposals in July at a meeting in which Ministers were asked to restrict to a minimum the scope of proposed legislation to allow sufficient time to pass the necessary Bills to dismantle the remaining emergency powers. Monckton acknowledged at this meeting the unlikelihood of passing legislation to effect the recommendations made in the Gowers Report and suggested that 'some administrative action' be taken to alleviate pressure from the TUC for the establishment of a national occupational health scheme.[61] Consequently, Monckton reconstituted the IHAC in October 1954 to advise on measures to further the development of industrial health services in workplaces covered by the Factories Act within the framework of the existing legislation, terms of reference which precluded any consideration of health problems in workplaces outside the remit of the Factories Act.

If this was an attempt at consensus politics, it was unsuccessful: neither the TUC nor employer organisations were placated. Writing to the secretary of the BEC, John Sadd of John Sadd and Sons, a timber import firm, complained that an investigation into industrial conditions was undesirable in an era of high employment, when British employers faced increasing pressure from overseas producers. 'Surely the time has arrived when the unproductive load on industry must be lightened and not increased', Sadd demanded:

> Surely genuinely unhealthy industrial working conditions must be the exception rather than the rule...In these days of over full employment if the people are not satisfied either with their rates of pay or working conditions, they can with the greatest of ease obtain

other work. Politicians painting pictures of impossibly high standards, involving as it does more committees, more snoopers, more unproductive personnel, more form filling, can only hasten the day of the country's economic strangulation.

Sadd's comments reflected a growing awareness that British industry was struggling, despite a period of economic growth and high employment in the post-war years. Difficulties experienced meeting target levels of armaments and munitions during the Second World War had brought the productivity of British industry under scrutiny. In the post-war years, governments explored how industrial efficiency might be enhanced and exports increased through greater state intervention, the proliferation of joint production committees, the modernisation of equipment, the professionalisation of managerial practice and the establishment of working parties to explore the difficulties experienced in particular industries.[62] Given the perilous financial state of the country after the War, successive governments viewed the modernisation of industry as imperative. As Sadd's letter indicates, however, many employers remained hostile to government intervention. The government's decision to pursue the question of industrial health services through the reconstituted IHAC appears to have been viewed by Sadd as yet another example of unnecessary government interference in industrial operations, and he took umbrage with the composition of the Committee:

Boffins from seven societies who have no interest in and no knowledge of industrial economics; representatives of the only people to benefit, and a representative from only one organisation which presumably has its feet on the floor... Altogether the set-up is unnecessary and badly conceived.[63]

The Confederation's Director George Pollock expressed his sympathies, noting that the Monckton's motivation in establishing such a committee was to prolong the opportunities for procrastination, 'to forestall criticism by the TUC that the "Dale" Report has been shelved. The TUC, you may remember, asked for something much more ambitious than is now being proposed, namely, what they called a "Comprehensive Medical Service".'[64]

At the general election in May 1955, the incumbent Conservative government increased their majority, thus averting the comprehensive occupational health service proposed by the Labour Party. In December of that year, Iain MacLeod, the erstwhile Minister of Health, replaced Walter Monckton at the Ministry of Labour. Given MacLeod's opposition while serving as Minister of Health to Walter Monckton's proposed national occupational health scheme, it is perhaps unsurprising that he did little to further occupational health services while in office. 'I have never accepted that ministerial responsibility for an industrial health service should be with this Ministry', noted MacLeod in 1956, 'and it is not to be assumed that I have changed my views with my Ministry.'[65] On the advice of the newly reconstituted IHAC, the Ministry of Labour instituted two pilot surveys. The first, undertaken in Halifax, investigated the conditions appertaining within factories in a locality.[66] This survey found that the factories in Halifax, which between them employed more than 22,000 workers, employed only four part-time doctors to provide a total of 15 hours a week to this work. Environmental conditions were classed as unsatisfactory in 25 per cent of workplaces while first aid arrangements were deemed unsatisfactory in 267 of the 760 workplaces surveyed. Eighty-one buildings were found to be structurally unsound.[67] These findings indicated that an extension of medical and nursing services would be advantageous and the Minister of Labour invited the managers of factories employing 250 or more persons to cooperate voluntarily to achieve this objective. Noting in 1960 that employers had not acted upon this invitation, Andrew Meiklejohn, Senior Lecturer in Industrial Health at Glasgow University argued that 'if industrial health services are to be developed in small factories, it will only be through legislation'.[68]

Trades Council members subsequently established a Halifax Industrial Health Committee through which they sought to expand medical services for workers, expressing particular enthusiasm for an industrial health centre staffed by an independent doctor. Frustrated by the lack of progress, the Committee issued a statement to the press in June 1961 that noted 'very little progress can be recorded...The Committee feels unanimously that industry should take the steps required in advance of legislation which will undoubtedly, in the not too distant future, make Industrial Health provisions obligatory.'[69] Faced with employer apathy and government inertia, the Committee approached the Nuffield Foundation for financial support in 1963, only to be informed that the Foundation's money had been committed to other schemes.[70]

A second survey, undertaken in Stoke-on-Trent between 1956 and 1958, focused on the health problems experienced within a particular industry, in this instance, the pottery trade.[71] This report was passed on to the Joint Standing Committee of the Pottery Industry, but Meiklejohn felt that the time for discussion had passed. 'We should no longer be discussing the need for the development of an industrial health service', he wrote, 'but only how, how soon and how much.'[72] Similar frustrations were expressed in the Commons in 1957 when the Minister of Labour procrastinated on the future of industrial health services. 'It is very heartening to read ... that much thinking is taking place about this matter', commented MP Mr Willey, 'but is it unrealistic to expect effective action in the near future as a result of that thinking?'[73]

A report compiled in 1957 by the TUC's Social Insurance and Industrial Welfare Department which outlined the existing provisions reiterated that the most desirable outcome would be a comprehensive health service.[74] By 1957, the Factory Inspectorate contained 18 medical inspectors, while 1,750 general practitioners worked on a part-time basis as appointed factory doctors to undertake medical examinations of young persons entering industrial employment. As regards private provision, 400 full-time and 3,000 part-time doctors were employed within factories. This represented a doubling of full-time personnel and a fourfold increase in the number of doctors employed on a part-time basis since 1948. Nevertheless, the Report critiqued existing provisions, in which less than 2 per cent of factories had definite arrangements for medical services and only 1 per cent provided a service which could be called comprehensive. It argued that the government needed to decide on the lines of future development so that a start could be made in building a comprehensive service. Subsequent surveys undertaken by the Ministry of Labour into the extent of voluntary medical supervision within factories in the 1960s would appear to vindicate these concerns. The number of full-time doctors working within factories had dropped slightly to 341 in 1965 and 339 in 1967 while the survey listed the number of part-time doctors correspondingly as 4,722 and 3,722.[75]

In 1960, the TUC continued to vent its frustration that little progress had been made towards the establishment of a comprehensive occupational health service. A 1959 congress resolution, carried unanimously, stated that 'immediate steps should be undertaken to establish a comprehensive Occupational Health Service', and asked for legislation to be introduced as soon as possible.[76] In the same year, the International Labour Conference passed recommendations for the establishment of occupational health services in all workplaces.[77] Following the general election in October 1959, Edward Heath replaced Iain MacLeod as Minister of Labour. Notes outlining the possible shape of a memorandum on medical supervision for the new Minister suggested that the choice lay between:

i) continuing over the next five years as over the previous, i.e. express-ing general good-will towards the idea of development, having at six-monthly intervals rather pointless discussions at the IHAC, on various minor measures which could be taken to stimulate develop-ments while continuing to avoid the main issue with its financial implications and the awkward problem of overlap with the National Health Service;
ii) having a new fundamental examination of the case for an occupa-tional health service and how to make significant progress.

While acknowledging that the collapse of grouped medical schemes for lack of funds might give rise to mounting criticism, the notes never-theless suggested that 'there is a lot to be said for (i) if the Government is prepared to face probably growing criticism from the "vested interests" on the IHAC, (e.g., TUC, BMA, and other "medical" groups) at lack of progress and lack of a real policy'.[78]

Putting its case to Heath, the TUC stressed its concern at 'the contin-ued lack of progress in developing such a service'. Like Meiklejohn, the TUC argued that responsibility for planning and developing services lay in the hands of the government as managers had 'neither the will not the ability' to initiate grouped medical schemes.[79] In January 1960, a memorandum on industrial health prepared for a meeting at Chequers noted that despite pressure from the TUC, the BMA and the majority of the members of the IHAC, the lack of progress on industrial health services did not arouse 'any general public interest'. Noting the lack of public demand for a government-run comprehensive occupational health service, the memorandum suggested that no 'embarrassment' would result from deciding against such a service in favour of a con-tinuation of the present policy of encouraging voluntary provisions.[80] In a statement given to the House of Commons, Heath made vague, non-committal statements about his intentions to 'explore the possi-bilities of establishing group schemes in areas of a different kind', and informed the Commons that the Nuffield Trust had decided to allocate a further £250,000 to develop group industrial health schemes which held out the potential to be self-supporting.[81] Critics pointed to the currently limited nature of grouped health schemes and queried how Heath expected these to expand to encompass all small firms unless directed by the government. 'I know that one school of thought would like to have a completely nationalised health service running parallel to the existing National Health Service', Heath responded. 'I take the view that this is a better way of approaching the problem.'[82]

This was not a view expressed in the 1962 Porritt Report, a review of Britain's medical services undertaken by a committee which represented a number of major medical organisations, and described by one Ministry of Labour official as 'a pretty poor effort'.[83] It advocated a hybrid occupational health service, run in association with or partly administered by the NHS through the Ministry of Health: recommendations which echoed the BMA's report on the future of occupational health services, published the previous year.[84] Articles published in the *British Medical Journal* on the future of occupational health services in 1950 and 1954 had stressed the need for further research into the forms such a service should take.[85] By 1961 the tone was decidedly different. 'For at least twenty years the British Medical Association and the Trades Union Congress have repeatedly urged the Government to take more initiative in developing occupational health services', noted a leading article. 'The cost has been a strong deterrent to action', it concluded, but 'most will agree with the BMA Council's opinion that this can no longer be a legitimate excuse for Government inertia.'[86]

Small factories and grouped schemes: a way forward?

Back in 1947, while the TUC pondered what responses to give to the IHAC Sub-Committee's questionnaire, Dr Schilling of the Association of Scientific Workers suggested that a different approach might help meet the health needs of workers. Schilling had recently visited the United States where he had investigated the provisions made to provide medical provisions to groups of small factories and felt it might be a useful experiment to try in Britain.[87] Subsequently, Schilling dispatched some notes on the industrial health unit in Long Island city, which had started to operate in January 1945.[88] A joint endeavour of the New York Health Department, State Department of Labour and Public Health Service, the scheme provided a service to seven factories with a total of 2,100 workers; the smallest factory in the scheme employed only 70 people. Staffed by a full-time doctor, seven part-time doctors and seven nurses, two engineers, a chemist, an industrial health expert and clerks, the service provided a model of the interdisciplinary industrial health team which would come to characterise developments in Britain.[89] Medical officers examined new employees to determine their suitability for different types of work, undertook examinations of workers returning from sickness, treated injuries and medical emergencies and were involved in the running of factory canteens. Medical supervision within individual factories was determined on a proportionate basis: for every 100 employees in the factory, a doctor spent three hours in the factory every week and a nurse spent nine hours. The works medical records were examined to see if any patterns could be determined in accident and absence rates. Other services provided by the industrial hygiene unit included an advisory service on industrial hygiene problems, periodic surveys of each factory to detect and control and industrial hazards, health education and a chest X-ray service for all workers.

A similar initiative was first attempted in Britain in 1947, when the Nuffield Foundation, Nuffield Provincial Hospitals Trust and Slough Estates Ltd collaborated to inaugurate the Slough Industrial Health Service. The scheme was financed by a per capita charge to member firms and grants from the regional hospital board and other sponsors. Initially, the scheme provided coverage to 42 firms with some 6,000 employees. By 1956, 178 firms had joined the scheme bringing a total of 17,000 workers within its provisions; at this stage, the scheme provided an emergency casualty service, rehabilitation facilities, a social service department, an occupational hygiene service and a records and statistical department.[90] The medical infrastructure was based around the central polyclinic, which was supported by two dressing stations to cover firms at some distance from the central clinic and a mobile dressing station. The large firms operated their own medical departments. Outpatient clinics were held in the same building as the central clinic, the Canadian Red Cross Memorial Hospital, which also housed facilities for X-ray, physiotherapy and chiropody.

The creation and maintenance of a healthy working environment was central to the scheme. A survey was undertaken of all firms when they joined the scheme and follow-up surveys were also carried out. Nursing staff visited the firms at monthly intervals and submitted reports on their findings. The occupational hygiene team investigated specific workplace hazards and provided advice on heating, lighting and ventilation. The individual health needs of the worker had not been lost in the attention to the environment, however. Indeed, the health services provided through the Slough scheme went some way towards establishing the model of a tailored service that would meet the needs of individual workers within and outside the workplace which had characterised the ideals of interwar industrial welfare supervisors and industrial psychologists. Rehabilitation facilities were sited at the Farnham Park centre, which was designed to assist short-term cases recovering from illness or injury, and offered physiotherapy and hydrotherapy, group remedial exercises and recreation; work of a therapeutic value was also available. The Social Service Department, staffed by a medico-social worker, was established in 1951 to deal with social, economic and domestic problems. The Department also assisted in the placement and follow-up of disabled workers.

The TUC's response to the scheme was initially hostile. When Mr Leyshon, a member of the Slough Trades Council met with Mr Owen of the TUC's Social Insurance Department, he expressed his belief that while continuing to support the idea of a state industrial health service, 'we ought not to reject without due consideration any scheme which would form a useful stopgap...individual services of this kind might be better than nothing at all'.[91] The TUC representative expressed concern that employers retained too much control over the scheme and could dictate its policy: there was little trade union representation in the council of management while representatives on the workers' councils had been selected by employers who had disregarded shop stewards and the unions. However, the main area of contention for the TUC was that the scheme might impede progress towards a national service. The TUC representative explained that 'whilst the TUC was prepared to look with a benevolent eye upon schemes which had as their object the health and welfare of those employed in industry, our policy was to be satisfied with nothing less than a state industrial health service'. As many potential interwar collaborators had similarly been informed, the Trades Council representatives were told that 'it would be treading on dangerous ground for the TUC to create the impression that official support is being given, in view of declared policy'. Ten years later, the TUC's approach towards the Slough scheme had moderated and Social Insurance Secretary C. R. Dale paid a visit to the services. By this stage a number of other grouped services had been established and the TUC had begun to consider whether, in fact, this approach might offer a realistic way forward, given that nearly half of all factories employing more than 250 people provided medical services, while only 1 per cent of factories employing fewer than 250 people did so.[92]

Since the end of war, the memorandum noted, the General Council of the TUC had been attempting to persuade the Factory Department of the Ministry of Labour to establish grouped schemes. Noting that the position remained substantially as it was in 1946 and that there seemed 'no likelihood that the Department will show any more readiness in the future than it has done in the past to sponsor developments of this kind', the memorandum argued that 'unless the trade unions themselves are prepared to take the initiative on the lines suggested in the attached memorandum, the prospects of making any real progress seems extremely remote'.[93] The rationale recalled the arguments used in the interwar years whenever the TUC had contemplated inaugurating a service or scheme to kick-start government involvement. 'By taking the lead to get a joint scheme of this kind going', the memorandum noted, 'the General Council would be making a real contribution in the field of industrial health.'

The foundry industry was suggested for a five-year pilot scheme as it was believed it would give good results. The dust, heat and heavy nature of the work made the trade very hazardous and caused injury and disease, therefore the provision of an industrial health service could be expected to yield prompt results in improving the health of the foundry workers covered. Undertaking this pilot scheme would provide the TUC with practical experience as to how grouped services might best be organised to meet the health needs of small workplaces, knowledge which could then be extended across industry.

The TUC hoped that by inaugurating such a scheme, it could pursue a number of objectives. A grouped scheme would not only provide treatment for injuries and ailments; the records gathered would help identify and ultimately prevent the occurrence of sickness and accidents directly related to the nature of the work. The scheme would focus on raising the standard of environmental conditions above the minimum level required by the Factories Acts and would investigate and control health hazards. It envisaged employing a full-time industrial medical officer and state registered nurse to run a service covering 3,000 to 4,000 workers within a prescribed area. Technical advice on the improvement of working conditions and the control of hazards would be sought from engineers and external bodies. These centralised facilities would be bolstered by the provision of a basic treatment area at each firm and the employment of auxiliary nursing and record keeping personnel. The TUC estimated that the running costs of the scheme would be between £4,000 and £5,000 and envisaged that the money could be elicited from trade unions, employers and the government.

However, the proposed grouped health scheme for foundry workers never made it beyond the drawing board; like other proposed TUC initiatives, it floundered when other organisations moved into the field. Nuffield funding proved decisive in the establishment of a number of grouped schemes, most of which operated on a similar basis.[94] In 1954, Nuffield funding facilitated the establishment of a grouped industrial health service at Harlow on an experimental basis. By 1962, the Harlow scheme boasted a central clinic, outpost clinic and mobile services. Like Slough, Harlow operated on a non-profit making basis; member firms paid a little over £2 per annum per employee, £1 if the firm has their own factory nurse. By 1962, 72 firms with a total of 11,000 employees had joined the scheme. The running of the scheme was overseen by a council of management which included representatives of management, members of the TUC General Council and the Harlow Trades Council.

Nuffield funding also underpinned the establishment of the Central Middlesex Occupational Health Service, established in 1956. By 1962, 53 companies had joined the scheme, providing cover to 7,000 workers. This scheme offered a casualty service for factories in the vicinity of a hospital, four clinics and an administrative centre. In common with the other Nuffield-funded schemes, the service was administered by a non-profit making company, with trades council representatives and employers sitting on the board of management alongside local medical interests. Smaller, Nuffield grant-aided schemes operated at Newcastle-upon-Tyne and Manchester, offering an advisory service to local firms and further schemes were in the process of being set-up at Rochdale, Dundee and District and West Bromwich, Smethwick and District Manufacturers.

Ronald Johnston and Arthur McIvor were doubtless correct in judging that the existence of the NHS dissuaded employers from investing in grouped schemes; nevertheless, their assertion that the schemes folded quickly once Nuffield funding was withdrawn is somewhat misleading.[95] Writing to defend the utility of grouped schemes in the face of government pessimism in 1980, F. H. Tyrer pointed to the longevity of established schemes in Slough, Harlow, Central Middlesex, Dundee, West Midlands, Rochdale, North of England and Milton Keynes, noting that only one which had operated in London had been forced to close.[96] Tyrer acknowledged the role played by Nuffield Trust funding in the establishment of such schemes (only the Milton Keynes scheme, financed by the Regional Health Authority and the New Town Development Corporation, had not been a beneficiary). He also attacked the role played by successive governments: 'periodic fanfares and the short-lived publicity attending the opening of a new service by some junior Minister have alternated with long periods of indifference', he wrote. 'Their experimental nature has been stressed, but no one has ever defined the terms of the experiment or the criteria on which eventual success would be judged.' 'Politicians', he judged, 'are notoriously ignorant about occupational health.'[97]

Some of the schemes had experienced setbacks; Rochdale, where Tyrer himself had been employed as Director of the Service, had been founded in 1962 with Nuffield funding. By 1969 the scheme was operating on a self-supporting basis and had its own premises and equipment. An economic downturn reduced membership of the scheme, which was forced to operate without a medical director from 1972, providing a mobile nursing service with part-time medical backing. At the end of 1978, 80 employers belonged to the scheme, bringing a total of 6,393 employees within its remit.[98] The Slough service also had to make cutbacks, relinquishing its rehabilitation centre, Farnham Park, to the NHS. It also had to surrender its industrial hygiene laboratory: the Minister of Labour refused to provide funds for its continuation on the grounds that all health needs within the workplace were adequately met by statutory regulations enforced by the Factory Inspectorate and that the government should not support voluntary facilities.[99]

The history of the grouped services demonstrates the marginal role played by the government in the provision of industrial health services. Indeed, Tyrer was perhaps not so far off the mark when he accused government officials of attacking and undermining these initiatives; he described how one former Chief Medical Inspector of Factories stated at an international conference that 75 per cent of the work done by the grouped services duplicated the work of the NHS.[100] However the continuance of many of these schemes in the face of government indifference, bordering on hostility, demonstrates that some employers were willing to invest in health services. Indeed, grouped occupational health schemes at Harlow, Milton Keynes and Slough are still operational alongside grouped services in Telford, East Anglia and London.[101]

Conclusion

The fate of industrial health services after the Second World War is not a straightforward story of decline: quantitatively, provisions actually increased. The Report created in 1957 by the TUC's Social Insurance and Industrial Welfare Department found that 400 full-time and 3,000 part-time doctors were employed within factories which represented a doubling of full-time personnel and a fourfold increase in the number of doctors employed on a part-time basis since 1948.[102] The establishment of diplomas of industrial health and university departments of industrial health afforded the previously marginalised discipline of industrial medicine a chance to escape its 'Cinderella' tag and go, if not to the ball, then in strength to the factory.[103] Moreover, creative solutions were being sought to provide cover to the smaller factories which predominated within the industrial sector with the establishment of grouped schemes, an idea imported from America. These provided a range of services, ranging from medical supervision to rehabilitation and environmental analysis. Underlying all these developments was private initiative and, most notably, a river of Nuffield Trust money. However, the government demonstrated no commitment to maintaining these provisions. Commenting on the uncertain future of the Occupational Health Unit at the Middlesex Hospital, which had been established with a grant from the Nuffield Trust, a lead article in the *Transactions of the Association of Industrial Medical Officers* asked:

> how can employers be expected to take seriously the declared aim of Government to encourage the development of occupational health services ... if it compels the closing of the only hospital-based service in the country by denying it funds – and this at a time when, we are told, the nation is enjoying unprecedented prosperity?

'It seems we have failed to teach our politicians that "prevention is better than cure" is more than platform cliché', the article concluded.[104] The Home Office Industrial Museum, which had epitomised the government's interest in health promotion within the workplace in the interwar years, was closed by the government in 1980 to save money.[105]

Action speaks louder than words, as the popular idiom would have it. However, a cursory examination of government initiative in the field of industrial health following the aftermath of the Second World War reveals a good deal of talk, but a demonstrable unwillingness on the part of the government to commit to a decisive course of action. It could be argued that the development of industrial health policy through a series of tripartite committees representing the interests of employers, trade unions and the State, rather than through new legislation, epitomised the politics of consensus which many historians have argued characterised the immediate post-Second World War era.[106] This approach did not produce any tangible results, however, and while *The Rise and Fall of the Healthy Factory* has not made the question of political consensus an explicit object of study, the conclusions arrived at would appear to chime with Nick Tiratsoo and Jim Tomlinson's assessment that 'dissonance and not harmony' characterised discussion on industrial policy.[107]

When pressed to take action, invariably the government would establish a committee to discuss matters further. 'Any number of committees may meet to discuss OH [occupational health] provision', Morris Greenberg recently commented, 'but their reports will moulder until there is a will for implementing their findings.'[108] The government lacked that will. When an interested party attempted to coerce the government to act by referring to the work of a previous committee, they were consistently informed that recent developments had rendered the findings of that particular committee obsolete and that more discussion was required. Indeed, ministerial files demonstrate that the government's explicit objective as regards occupational health services was to drag its feet for as long as possible while minimising the political fallout of such an approach. Small wonder, then, that actors from across the political spectrum decried the government's strategy of procrastination and expressed palpable frustration.

Industrial medical services did not, then, disappear. A more accurate description would be that expectations as to what an industrial medical service should consist of and what role the State would play in such a service were radically transformed between 1945, when a state national industrial health service seemed imminent, and 1960, when Heath procrastinated once more on the future of services in the Commons.

The fate of industrial health provisions throughout the period under study remained inexorably linked to the economic fortunes of the country and this remained very much the case in the post-war era. In this period, the problem was not the burden of unemployment and depression which had faced workers and employers in the interwar years, but a shortage of labour and a growing anxiety that international competition and industrial inefficiency would damage British industry. Returning to the start point of this study, concerns that Britain's industrial manufacturing sector was underperforming had inspired the government to establish the HMWC. In short, the new ideal of the healthy factory had been fundamentally rooted in efforts to enhance industrial and national efficiency.[109] By the 1950s, efficiency was once more a key issue but no longer acted as a driving force for innovative health services. The belief that the efficiency of the industrial machine was directly related to the mental and physical health and efficiency of the individual worker had disappeared. The individual in the workforce had been lost and the boundaries between home and workplace once more severed. Factories were no longer sites of health production; they had become once more simply sites of economic production. In this new context, health services were actually construed as counterproductive to the goal of raising industrial efficiency and competitiveness. One might speculate as to whether a Labour government would have inaugurated a comprehensive occupational health scheme had it not been kept out of power by the Conservatives between 1951 and 1965.

Pertinent questions still remain, however. For example, did the TUC err in its decision to pursue the Ministry of Labour as the appropriate department to develop industrial health services or would it have stood more chance had it backed the Ministry of Health? Moreover, while provision may have increased in quantitative terms, the concept of an industrial health service, and more broadly the ideal of the healthy factory, appears to have undergone a radical transformation in this period. To explore these matters further and to study in more detail the impact of the NHS on proposals for a state industrial health service, the final chapter will investigate the evidence given by different organisations to the 1948 IHSC.

6
The Fall of the Healthy Factory

Towards the Dale Committee: industrial health, medicine or hygiene?

A flyer issued by the Department of Occupational Health at Manchester University to publicise a conference held in 1951 on 'The Role of Industrial Medicine in the Welfare State' claimed that training for industrial medical officers had become caught in a time warp, governed by a concept of industrial medicine which had crystallised by 1942 at the height of the wartime expansion of industrial medical services within factories.[1] The purpose of the conference was to ascertain the purpose of industrial medicine now that the NHS had been established and it raised a number of questions about the purpose of industrial medicine and the role of the medical practitioner within the factory. Only recently, it claimed,

> Did many of us become aware of those disturbing questions "Am I really necessary?" "Am I doing a job which no other professional worker can do as well or better?" "Why am I here, an industrial medical officer or nurse, in a factory?"

The expansion of medical personnel with factories during the Second World War at the instigation of the government gave rise to the belief that a state industrial medical service would be inaugurated following the cessation of the war in tandem with a national health service. Even during the course of the Second World War, however, some government officials, particularly those based within the Factory Department of the Ministry of Labour proposed a different vision of how an industrial medical service might work. The Chief Medical Inspector of Factories, Edward Merewether, promoted the idea of an industrial health as opposed to an industrial medical service, in which engineers, chemists, personnel officers and dieticians would work alongside doctors. Industrial health problems, he asserted, were 'preponderantly environmental'.[2] In the previous chapter, the persistent endeavours of the trade union movement to secure a form of national industrial health service following the exclusion of industrial health provisions from the NHS were explored. This chapter helps explain in part how and why other interest groups were able to resist these demands by moving the goalposts as to what constituted an industrial health service, painting TUC demands for medical personnel as outmoded and out of touch with real needs. Consequently consensus as to the purpose and form of industrial medicine was eroded over the course of the 1940s, prompting the professional crisis of confidence reflected in the conference organised at Manchester.

In 1948, the Ministry of Labour and National Service sent a statement to the TUC which outlined government policy on the issue of industrial medical services within factories, considering the place of any proposed industrial medical services in relation to the recently established NHS.[3] This latter service, the statement claimed, now provided a 'full range of medical treatment' and 'ordinary personal health doctoring for individual patients'. Although acknowledging that doctors working within industry were to some extent concerned with the personal health problems of individuals, the government asserted that industrial medical officers primarily had to 'consider the health problems of workers as members of groups'.

Industrial health problems, 'whether of a general kind or at particular establishments', were described in the government's statement as 'largely non-medical and largely beyond the scope of any individual doctor', falling more within the remit of laymen such as engineers, chemists and managers than doctors. Contrasting the role of the doctor engaged in industry with the doctor employed by the NHS, the statement envisaged that the industrial doctor, trained in industrial conditions, would work closely with specialist laymen as members of an industrial health team. In his history of the Association of Industrial Medical Officers, J. T. Carter explored how the scope of industrial medicine was redefined following the introduction of the NHS, as earlier beliefs that the workplace could act as a base for the general practice of medicine were 'severely curtailed'. Carter traced how some industrial medical officers responded to the change in affairs wrought by the implementation of the NHS by retreating into clinical practice while others found their role supplanted by the technical skills brought into the workplace by the new members of the interdisciplinary team who were more ably equipped to contain and control health risks.[4]

The government's statement also questioned how practical it was to extend medical supervision to encompass small factories. Of the 230,000 factories within Britain, around 195,000 employed 25 people or less. The statement claimed that such factories posed no special health risks beyond cleanliness, temperature, ventilation, lighting and welfare facilities and that infrequent visits by doctors to such places would be of limited value. A substantial increase in medical manpower would be required if the intention was to provide all factories with comprehensive medical supervision and, noting the shortage of doctors with specialised qualifications and training in industrial medicine, the government claimed that this model did not appear to be 'practical politics' in the immediate future.

Readers were informed that 'the policy of the government is that there should be a national industrial medical service properly co-ordinated with the NHS', but only 'as soon as there is a sufficiently large cadre of suitably qualified doctors to form a sound foundation for the establishment of such a service'. In the meantime, it claimed, steps should 'continue to be taken' to encourage the building up of such a cadre. This statement of policy must have been frustrating reading for the TUC which for a number of years had been urging medical schools to develop specialised diplomas and to place a greater emphasis on industrial health issues within the undergraduate medical curriculum, with little success.[5] It clearly bore the hallmarks of Edward Merewether, replicating many of the remarks he had made in his role as Chief Medical Factory Inspector during the Second World War. By stressing the 'specialised needs' of the industrial medical service, the government in effect emphasised the distinction between industrial health services and the future NHS which it had established in the publication of the White Paper on the NHS in 1944.[6]

While this statement depicted a cabinet united as regards the shape of future government policy for occupational health services, letters exchanged between the Ministries of Labour and Health paint a rather different picture. In December 1948, the Socialisation of Industries Committee invited the Minister of Health and the Secretary of State for Scotland to present a memorandum on the provision of medical services in socialised industries.[7] The resulting joint memoranda portrayed industrial medical services as a rampant weed which urgently required judicious pruning. Both ministers voiced their concern regarding 'the growing threat that money and manpower may be wasted through the un-coordinated development of health and medical services' and viewed the 'ambitious' plans to expand medical provisions in the social-ised industries as 'alarming'. Advocates of industrial medical services, in their view, had failed to take into account the advent of the National Health Service and they lambasted the protocols established by the Minister of Labour to monitor industrial health developments – namely the Industrial Health Advisory Committee – for failing 'to check the growing and wasteful demand on trained manpower for the socialised industries'. The memorandum concluded that an Order in Council be issued under the Ministers of the Crown (Transfer of Powers) Act of 1946 to transfer all responsibilities for industrial health services to the Minister of Health and the Secretary of State for Scotland, who would then consider what the scope and nature of health services for industry should be.[8] George Isaacs, the Minister of Labour and National Service, subsequently wrote to the prime minister, Foreign Secretary, Secretary of State for Scotland, Minister of Health, Minister of Supply and Minister of Fuel and Power to 'express my great surprise that a proposal to take over part of my functions should have been circulated without inform-ing me beforehand', insisting that the matter be dealt with through the Official Committee on the Machinery of Government.[9] Following a fur-ther meeting of the Socialisation of Industries Committee which con-sidered memoranda submitted by the ministries of Health and Labour, the Lord President of the Council and the Chancellor of the Exchequer recommended that a committee be set up to investigate the matter.[10]

Consequently, in June 1949, the Prime Minister announced the establishment of a committee to investigate the relationship and any potential overlap between the preventive and curative health services provided for the population at large and the industrial health services.[11] The statement stressed the need to make the best use of the country's limited medical resources and the necessity to avoid any duplication of service. The government requested that companies postpone any fur-ther development of their medical services until the Industrial Health Services Committee (hereafter IHSC) had reported, which the Prime Minister anticipated 'will not take very long'.[12] From the outset, ques-tions were raised as to what such a committee would be able to accom-plish. Given that the government had been under significant pressure from the TUC and medical organisations to develop an industrial medi-cal service, it is quite striking that the basic premise underlying the establishment of the IHSC was that industrial health services robbed the NHS of manpower. Examining the claims made in evidence given to the IHSC by the Ministries of Health and Labour, the British Employers' Confederation (hereafter the BEC) and TUC, it becomes apparent how widely concepts of industrial medicine had diverged.

Industrial health services: a threat to the NHS?

One of the most striking claims made in the evidence given by the Ministry of Health, an allegation reflected in the remit of the IHSC, was that industrial health services had become overdeveloped. The Ministry's written evidence questioned how much further industrial health services could be expanded 'without robbing those general services of manpower', claiming there was evidence of a tendency for industry to attract doctors away from public health work.[13] Nursing levels were also stated to be under threat: 'There are some 4,000 state registered and other nurses employed in industry', the Minister noted, 'at a time when some 50,000 beds in hospitals are closed for lack of nursing staff.' The Ministry asserted that the tendency was for industrial health services to grow, and, 'if one pushes that to its logical conclusion, one gets to a point when all employers, big and small, will have doctors and nurses to attend to their own employees.' If parallel development was allowed to continue, 'every factory would have its medical officers and its medical nurses, and the general health services would gradually have their medical personnel drained away'.[14]

But was this truly the case? Employer organisations were very dismissive of this argument. At a meeting of the Factories Committee of the BEC, committee members stated that they 'doubted whether there was any case where existing industrial health services could be considered excessive', as 'no employer was going to pay a doctor or nurses to do little or nothing'. Indeed, they admitted that medical coverage within smaller factories was likely to be deficient.[15] Members of the BEC made a number of suggestions as regards ideal levels of industrial medical provision; one respondent urged all factories employing more than 1,000 to institute an industrial health service,[16] while another considered the employment of a state registered nurse desirable in factories employing more than 250 people.[17] However, the Confederation rejected as ridiculous a suggestion put to them when giving evidence to the IHSC that a further 25,000 doctors would be required to run an industrial medical service. 'If you are saying that is a perfect state', the BEC's representatives responded, 'then everybody in their homes should have a doctor sitting on their heads all the time.'[18] Admittedly the NHS was facing a nursing staff shortage in its early years; concern at the shortage of trainee nurses prompted the government to establish the Working Party on the Recruitment and Training of Nurses in 1946, otherwise known as the Wood Committee.[19] However, developments since the cessation of the Second World War provided no grounds for the Ministry of Health's hypothesis that industrial health services would continue to expand if unchecked: the BMA noted that in 1948 there were 200 full-time industrial medical officers and 700 part-time;[20] the respective figures in 1944 had been 180 and 890. Moreover, the number of nurses working in industry had halved from over 8,000 in 1943 to 4,000 by 1947, of which only 2,600 were state registered.[21]

When Ministry of Health representatives gave oral evidence to the Committee, their assertion that industrial medical services robbed the NHS of manpower was repeatedly challenged by Sir Geoffrey Vickers, legal adviser to the National Coal Board.[22] The Ministry of Health argued that industrial health services should be taken over by the NHS and run through local authorities by local medical officers of health. It claimed that this plan would economise on medical manpower. However, under cross-examination, the Ministry's representatives admitted they had not planned how an industrial medical service would work in detail, were unaware what percentage of workers received any form of industrial medical services and did not know how many doctors would be needed to run the kind of industrial medical services which they envisaged.[23]

Vickers asserted that only around 5 per cent of workers received any form of industrial medical service and he queried how the other 95 per cent of workers would be covered without some expansion of personnel. 'There are something like nineteen million persons in employment', he stated, 'so that even your 4,000 nurses would not go far.' He ridiculed the suggestion that existing personnel within industrial medical services could be redeployed to provide coverage to all workplaces, and still have staff left over to channel back into the NHS. 'Are you suggesting that if a comprehensive and country-wide industrial medical service was established on the lines you suggest it could spread the butter so thin as to reduce rather than increase the number of medical and nursing man-hours?' he asked. 'If so, I should have thought that the butter would have been almost invisible.' Continuing his food analogy, Vickers asserted that, 'You could not feed all the industrial workers from the crumbs that fell from the table of the few who now have an industrial health service.'[24]

Another way to explore the Ministry of Health's assertion that industrial health services were overdeveloped at the expense of the NHS is to calculate what percentage of the nurses and doctors working within Britain were employed within industry. It was widely accepted, for example, that around 4,000 nurses were employed within industry by the late 1940s.[25] In 1949, the BMA noted that there were 180,000 nurses within Britain; the number working in industry therefore represented just over 2 per cent of that figure.[26] In the minutes of one meeting of the IHSC, attention was drawn to the fact that while there were around 240 full-time doctors working for industrial health services, 30,000 doctors were employed within the NHS and a total of 55,000 doctors worked within Britain.[27] It was often pointed out that there was a larger body of industrial medical officers working on a part-time basis. However, a survey undertaken by the Ministry of Labour and National Service at the behest of the Industrial Health Advisory Committee in 1949 found that of the 1,021 doctors employed part-time, 504 worked fewer than three hours a week while a further 406 worked between three and 12 hours a week.[28]

What was industrial about industrial medicine?

A further line of argument advanced by the Ministry of Health in the evidence it gave to the IHSC was that industrial workers had no health needs which could not be met by the general health care services offered by the NHS. Therefore, the argument ran, industrial health services which had developed in the absence of state health care had now been rendered redundant following the inauguration of the NHS.

When giving oral evidence, it was suggested to the Ministry of Health that the NHS did not meet all the needs of workers regarding treatment and diagnosis, lacking the expertise to deal with occupational diseases. 'In certain factories and workshops rather specialised conditions... require specialised services', the Ministry's representatives conceded. Nevertheless, they continued, 'I think the first thing to do... is to put that special problem on one side and look at that again later on. After all, it can only be a small proportion of the working population which is subjected to special occupational risks.'[29] This assertion was challenged by one committee member, who asked, 'Is not the Industrial Health Service really a specialised service?' And while the Ministry's representatives acknowledged that this may have originally been the case, they claimed it had now evolved into a service whereby doctors 'look after the general health of the ordinary workers in an ordinary factory and he looks after the colds and cuts and their rheumatism and their general physical condition'. The industrial medical staff working in a large modern factory, it was asserted,

> are giving what you might call a luxury service to their employees, whereas there are other people who are really sick and badly in need of attention from these people. It is true, to the extent that the doctors and nurses in that big factory are giving these luxury services, giving very full attention to coughs and colds and other mild disabilities, they are saving the NHS to that extent.

Much of the work undertaken by an industrial medical service, they claimed, was 'the administration of aspirins and sedatives and other harmless palliatives and some light dressings', activities which did not require the professional skills of a state registered nurse.[30] The Ministry's representatives gave the example of an engineering factory one had visited at Cardiff during the war in which the health service provided were described, once again, as of a 'general' character. The factory provided physiotherapy for its employees and the Ministry's representatives suggested that treatment was given not to meet medical needs but to boost morale. 'They had men standing round the walls of the room getting sunlight treatment and so on', the representatives claimed.[31]

The Ministry of Health representatives painted a picture of luxurious over-elaborate and over-staffed first aid departments, cluttered with light and radiant heat treatments and provisions for dental and eye treatment, all provided separately and competitively by employers, rather than through the NHS. The factory was reconceptualised as a mundane environment; industrial medical services met general health needs.

Ministry of Labour staff were highly critical of the Ministry of Health's evidence, deeming it 'tendentious' and 'very misleading'.[32] Edward Merewether, the Senior Medical Inspector of Factories, suggested that the picture of industrial health services presented by the Ministry of Health to the Committee has been gleaned from knowledge of the Ministry of Supply's wartime services, 'a small altruistic engineering firm at Cardiff' and the Slough Industrial Health Service. In Merewether's view, the Ministry of Health was obsessed with the extent of personal health services in industrial health services and failed 'to recognise the extent of the occupational environmental element provided'.[33] Given prior agreement that the Committee would not consider the division of departmental responsibility, the assertions made by the Ministry of Health that it should assume responsibility for industrial health services also provoked ire.

When giving oral evidence to the IHSC, representatives for the Ministry of Labour and National Service emphasised the industrial character of industrial health services, stressing the role of personnel management, welfare, and environmental factors such as ventilation and temperature. They argued that these aspects of industrial health services required expert oversight, could not be devolved to local authorities as the Ministry of Health had suggested and urged the expansion of industrial health services as a means of reducing the workload of the NHS.[34] Subsequently, the Ministry of Labour provided the IHSC with the results of an investigation it had carried out to try and ascertain what proportion of factories carried special risks of injury to health associated with the nature of an industrial process, material or plant used or kind of work done in a factory, as distinct from general health risks described by the Ministry of Health.[35]

Analysing thousands of trades carried on in factories and the number of factories in which the various trades were carried on, the Ministry divided trades into two groups: those in which special industrial health risks hardly ever or never arose, and those in which a visiting inspector would as a matter of course look into the question of whether a special health hazard known to be associated with that trade was in fact arising in the case of that particular factory. It estimated that about three quarters of factories in the country fell into the latter category. Factories, in its view, were by no stretch of the imagination 'ordinary' environments; the health risks they posed were specific and not general.

The BEC also stressed that industrial health services were best staffed by full-time doctors and state registered nurses holding industrial medicine qualifications. Allowing the local authorities to take over the functions of industrial health services as the Ministry of Health proposed would, the BEC asserted, result in 'a very great deal of inefficiency'. One of the greatest values of industrial medical services, the BEC's representatives claimed, was 'the knowledge of industrial risks, and also ... knowing intimately the works and the workpeople, and the conditions under which they are working'. Extending the range of activities given to first aid personnel within industry would be 'very dangerous indeed'.[36] This line of argument ran counter to claims made by the Ministry of Health that workplace health needs could be met by a partly trained first aider armed with some plasters and a box of aspirin.

A service to promote industrial efficiency or to protect the health of workers as citizens?

During the proceedings of the IHSC, the BEC argued that the main function of industrial medical services was not to provide a health service as such but to promote industrial output, and consequently that industrial workers should receive preferential treatment over the population at large. This latter point was made very clearly when the BEC representative on the IHSC, Mr Pheazey, put the following question to the Ministry of Health representatives:

> We are an industrial nation and everything, for all of us, depends on the worker, the people who are engaged in industry... Does that not seem to indicate to you that since the non-working man, woman and child are dependent upon the working man perhaps the Industrial Health Service should take first place and be ahead of the NHS?[37]

The Ministry of Health's representatives attempted to paint this as an outrageous suggestion by contrasting the needs of workers with the needs of mothers and children. 'Would you suggest that the care of the children of the country and of the mothers of the country was something that any Ministry could say must take second place even to the health of industrial workers?' they asked.[38] Nevertheless, a historical precedent to prioritise the health needs of industrial workers in order to maintain industrial production had certainly characterised government initiatives earlier in the century, as evidenced by the introduction of the National Health Insurance Act and government intervention in workplace conditions during both World Wars. When the BEC had solicited feedback from its members regarding industrial health services, one member of the Coventry and District Engineering Employers' Association indignantly complained that prior to implementation of the NHS industrial accidents were prioritised but now such injured workers 'have to take their turn with all National Health subscribers'.[39]

The assertion that workers should receive preferential services supported the BEC's broader claim that the main purpose of industrial health services was to promote industrial efficiency. 'The healthier the man, the more efficient he is, just as the better the machine, the better its output', stated the Coventry and District Engineering Employers' Association.[40] The implication of this perspective was that industrial health services did not belong within the remit of the Ministry of

Health. As one committee member suggested to the Ministry of Health representatives:

> The Industrial Health Services as set up at the moment are concerned not so much with the health of the individual as with the efficiency of the factory...why should either the Exchequer or the rates be responsible for a contribution which is entirely selfish, from the point of view that it is only concerned with efficiency?[41]

Certainly the BEC stressed how industrial health services benefited industry, enabling workers to secure immediate treatment for injuries and sickness. Without this facility, employers would lose many productive hours as workers waited at general practitioner surgeries to be treated, and workers would lose wages. The Coventry and District Engineering Association argued that a visit to the works medical centre would take at most an hour, while a visit to a hospital or general practitioner would take half a day: they calculated that 1,500 visits to the works medical centre entailed a minimum saving of 5,250 production hours per week.[42]

It was put to the TUC representatives when they gave oral evidence to the IHSC that 'the objective of the industrial health service is to keep the machine running to turn out the greatest number of goods and so forth, to throw the individual into danger, but to protect him' and that consequently industrial health services were 'an industrial matter, not a health matter'. They disagreed. 'The effect of both the NHS and the industrial health service is to keep the individual healthy', they argued. 'It does not matter for what purpose we are keeping him healthy.'[43] Three years later in an article in *The Transactions of the Association of Industrial Medical Officers*, the TUC's Social Insurance Secretary C. R. Dale continued to espouse the same perspective. 'Is our aim efficient production, or healthy men and women?' he asked his readers. 'The health of the men and women in industry is the concern of all of us', he asserted, 'whatever the special responsibilities of management', and consequently 'the fundamental unit of industrial health should be the individual'.[44] For the TUC, people at work were citizens and any national health service which failed to extend into the workplace to protect against the elevated hazards posed to citizens in this arena by potentially hazardous substances, practices and environments, was clearly remiss. However, other organisations were able to argue that people at work were undifferentiated workers first and citizens second. Health needs could be met outside the factory in services established for the population at large; any industrial health service which did exist did so to minimise inefficiency. This line of argument sought to re-establish the boundaries between home and workplace which had dissolved in the interwar years.

As these responses indicate, much of the debate sought to determine which government ministry should be responsible for the industrial medical services. By understating the risk to health and the industrial nature of that risk, the Ministry of Health was able to argue that industrial medical services should be incorporated within the NHS and that employers should be asked to maintain their standstill on industrial developments. This perspective was countered by the Ministry of Labour. It stressed the industrial character of many of the health problems arising in workplaces, even in small factories. The Ministry of Labour had tipped the balance the other way, arguing that industrial work posed a real risk to workers' health, but it claimed that the solution lay less in medical supervision and more in the intervention of other experts.

Other parties were very much aware that interdepartmental rivalry was at work. At a meeting of the BEC's Factories Standing Committee, one member stated:

> We knew that Mr Aneurin Bevan and Sir Wilson Jameson [Chief Medical Officer] were doing their utmost to secure that all Industrial Health Services should be under the Ministry of Health. It was fortunate, however, both that the Chairman, Judge Dale, was impartial and that the Factory Department of the Ministry of Labour was in profound disagreement with the Ministry of Health's views on this Question.[45]

While the BEC favoured a specialised service and therefore believed that the service should be managed by the Ministry of Labour, its main priority was to ensure that state intervention remained at a minimum. It argued therefore that that the industrial medical service sought foremost to promote industrial success.

An industrial or an occupational health service?

At a meeting in 1945 with Ernest Bevin, Minister of Labour, a TUC deputation discussed the feasibility of revising the Factories Act to extend provisions such as canteens, sanitation, mess room facilities and washing accommodation, which were covered under the Factories Acts, to all industrial and commercial undertakings. Bevin responded that he 'had always taken the view that there was more ill health in offices and shops than in factories' and suggested that the TUC produce a memorandum. While the interwar years had witnessed an explosion of interest in the design of factory buildings, Bevin claimed that office buildings were often improvised, lacking space and basic amenities such as washing facilities.[46] A number of organisations demonstrated similar thinking in the evidence they submitted to the IHSC, using the term 'occupational' in preference to 'industrial' when outlining their vision of health services for workers.[47]

The dawning recognition that the working conditions of a large proportion of the working population were largely unregulated led to the establishment of the Gowers Committee in 1946. Tasked to investigate the health, safety and welfare of people employed in workplaces which lay outside the scope of the Factories or Mines and Quarries Acts, the Committee reported back in 1949.[48] The Committee's recommendations for alterations to the law were largely modelled on provisions contained within the 1937 Factories Act regarding sanitary accommodation, space, ventilation and temperature, underground workrooms, cleanliness, washing facilities, accommodation for clothing, provision of seats, dangerous machinery, rest rooms, escape from fire and the provision of first aid boxes. The 1937 Act was viewed by the Committee as 'an important piece of modern legislation by means of which fairly detailed standards of safety and welfare have been brought into the working lives of a large section of the community', much of which it believed could be usefully adapted to meet the needs found within non-industrial employment.[49] While making these recommendations, the Committee sounded a note of caution. 'The gravity of Britain's economic difficulties during the years that we have been sitting has imported a certain unreality into our proceedings', the Committee wrote in an introductory setting entitled 'Economic Setting'. They elaborated further:

> We are considering conditions of employment...at a time when the labour and materials necessary to improve them are not sufficiently available. We are considering proposals for shortening the hours of work of juveniles at a time when industry must devote all its energies to increasing output. We are considering the need for fresh legislation that could only be enforced by new or augmented inspectorates at a time when every available man and woman is wanted in productive employment. Some witnesses have gone so far as to deprecate in present circumstances even the preparation of plans for the future.

The Committee proceeded nevertheless to produce its suggestions, while acknowledging that this economic context 'might make some of our recommendations not immediately practicable'.[50] It would take a further 14 years before the government enacted these recommendations in the 1963 Offices, Shops and Railways Premises Act.

The campaign for legislation to improve working conditions within factories had been premised on the grounds that the factory environment posed particular risks to health not present in other working environments; it had often been constructed as unnatural, depriving workers of fresh air and daylight and subjecting them to work with toxic substances. This had typically been compared unfavourably to the health-giving properties of agricultural labour. Thus, the Factory Inspectorate reported favourably on the installation of an open air workroom at one factory in 1913 which was intended to 'afford exceptional opportunities for regaining health to girls who are recovering from illness or who have markedly fallen below the standard'.[51] Similarly, office and shop work had traditionally been perceived as cleaner and less hazardous, though a resolution passed at the 1913 Annual Trades Union Congress indicates that unions at least felt there were potential health risks lurking in office work. Stressing the prevalence of consumption and other pulmonary diseases among clerks, the resolution argued that offices should be brought within the scope of the Factory Acts through an Office Regulation Bill.[52] These lines of argument gained force among the trade union movement in the late 1930s; one speaker at the 1938 Annual Conference of Unions Catering for Women Workers asserted that 'in many offices, the conditions under which young women worked were far worse than conditions in factories. They were crowded together, often in basements which were insufficiently illuminated.'[53] At the 1945 conference, delegates passed a resolution on office conditions which urged the TUC to give its support to the Offices Regulations Bill to ensure that the health of office workers was protected in light of the increase in tuberculosis and related diseases. In discussion, delegates expressed their belief that office girls were disadvantaged when compared to factory girls as they had no specific legislation to protect them.[54]

Similarly, by 1950 agricultural labour was being contrasted unfavourably to conditions appertaining within factories, some 60 years after the term 'factory farming' was first enthusiastically coined to describe how the application of factory techniques to farming practice could enhance agricultural output.[55] It was reported to the Annual Women's Conference that year that the TUC's General Council had been considering welfare facilities for agricultural workers and were urging the supply of clean water, soap and towels, first aid boxes and protective clothing for dangerous substances; provisions which had been available to factory workers for many years.[56] Passing a motion which called on the government to implement without delay the recommendations of the Gowers Committee regarding the provision of sanitary accommodation, washing facilities, safeguarding of machinery and protective clothing for workers using chemicals and other dangerous substances, delegates stressed the dangers of agricultural work. Miss Hymes, representing the National Union of Agricultural Workers, discussed the 'deadly nature' of some of the insecticides, weed killers and chemical sprays which many agricultural workers had to mix and apply. She described how workers came home with blistered and stained skin and reported that some washes now used killed all insect and bird life in the area applied. 'Why could not washing facilities be provided for farm workers as for workers in factories and mines?' she asked.[57] The campaign for greater regulation of agriculture in part sought to establish the same welfare facilities which had come to be associated with the healthy factory. In part also, the hazards posed to agricultural workers had been brought about by the application of industrial technology to farming practice, particularly the deployment of potentially dangerous machinery and toxic pesticides.[58]

Conclusion

Using the evidence given by different organisations to the 1949 IHSC,[59] this chapter has explored the decline of the ideal of health within industry, analysing how the terminology used began to shift from industrial health to industrial medicine and then to industrial hygiene. While an industrial health service carried connotations of a preventive service which could positively improve health, an industrial medical service was often evoked when a curative service was envisaged that would patch up injuries. Increasingly, both these terms were discarded and an industrial hygiene service was recommended which would employ technical experts such as engineers to ensure the health of the environment, with medical interventions and medical personnel taking second place. The receding interest displayed in the health needs of individual workers, discussed in Chapter 4, went hand in hand with these developments. These competing conceptualisations of what a service should look like shifted the debate away from the concept of the healthy factory which had emerged in the First World War and flourished in the interwar years. It also offered organisations keen to postpone and shelve the prospect of a state industrial health service grounds to suggest that further discussion needed to take place before action could be implemented, a phenomenon illustrated in the previous chapter. Organisations which clung more tenaciously to the concept of a health or medical service found that the parameters of the debate had shifted and were chided for failing to keep pace. 'It is not just a matter of attaching doctors and nurses to factories', the TUC were rebuked in a 1948 government statement.[60]

In the proceedings of the IHSC, a number of questions emerged which polarised opinion as to the nature and purpose of an industrial medical service. One crucial issue of contention was whether the primary objective of an industrial health service was to protect the health of the individual at work or to promote industrial productivity. Some sought to contend that the two objectives were inexorably linked, as they undoubtedly were to some degree, but the decision as to where the emphasis lay then impacted on other questions. If the treatment provided by industrial medical services was aimed at people constructed as productive wage earners rather than citizens, it could logically be construed that such services belonged to the sphere of industrial organisation and were a matter for employers and the Ministry of Labour rather than the NHS and the Ministry of Health.

A further question was the degree to which the health needs of workers engaged in industry required specialised services. Participants in the debate contested the extent to which workers' health needs centred on the place of work and the interaction of work and worker, and consequently whether these health needs could only be met by a specialised unit of health service. If it was agreed that specialised provisions were required, the question was then once more that of the politics of responsibility for sickness and treatment: would such a unit best perform its functions from within the framework of the NHS or from within the framework of economic and industrial units and should the responsibility for funding such a service rest at the feet of the State, employers or employees?

It was widely acknowledged that the fate of industrial health services was in part undermined by political rivalries between the Ministries of Health and Labour. While both ministries vied for ownership and control over industrial health, paradoxically neither actually wanted to accept the responsibility to develop the service. A 1948 TUC memorandum reporting on the work of the Medical Occupational Committee of the BMA noted that there were only 200 full-time and 700 part-time industrial medical officers, a statistic recognised by Sir George Schuster as 'appallingly low', yet 'the Ministry of Health disavows any responsibility for this service and have placed it on the Ministry of Labour'.[61] Indeed, the evidence given by the Ministry of Health to the IHSC indicates clearly that its attempt to purloin industrial health services from the Factory Department of the Ministry of Labour was motivated by a desire to impoverish the services further in order to bolster the NHS. The failure of any one government department to take responsibility for the development of industrial health services was perceived by the TUC and medical organisations as a primary cause for its retarded development. By 1957 the TUC had begun to question whether they had backed the wrong horse by championing the Ministry of Labour rather than the Ministry of Health as the rightful department to develop an industrial health service; the intentions of the Ministry of Health as stated before the IHSC indicate however that this would have resulted in an even less favourable outcome.

After 1946, all debates as to the future shape of industrial health services considered the fate of these services in relation to the NHS. This comparative analysis proved fatal to industrial health. On the one hand, the establishment of what was claimed to be a comprehensive health service dulled enthusiasm for any significant extension of industrial medical services: if the establishment of health care provisions within the workplace had acted as a disincentive to the provision of a system of national health care in America,[62] the reverse scenario was true in Britain. While the previous chapter demonstrated an increase in medical personnel employed within factories between 1945 and 1960 and the tenacity of grouped medical services, these provisions still only covered a fraction of the workforce. Employers did not withdraw services as such; the evidence collated by the BEC from its member organisations as it prepared to give evidence to the IHSC indicated that those employers who had established medical services were resistant to the suggestion that these be surrendered to provide personnel for the NHS. Nevertheless, there appeared to be little appetite among employers generally to provide extensive facilities at their own expense when the NHS existed to treat their sick and injured employees. While services remained, they were no longer characterised by the ideals of health promotion and preventive care that had come to denote the interwar model of the healthy factory. Researching the history of occupational health and medicine in Scotland between 1930 and 1974, Arthur McIvor and Ronald Johnston have arrived at similar conclusions, arguing that the establishment of the NHS led to the marginalisation of occupational medicine and the persistence of variable workplace health care provisions. They suggest that the opportunity to adopt a preventive health care approach at the workplace was largely discarded in favour of curative or palliative treatment through the NHS.[63]

The remit of the IHSC explicitly framed industrial health services as a threat to the NHS and this chapter has sought to demonstrate the absurdity of the suggestion that the tiny handful of staff employed by industrial medical schemes were responsible for the NHS staffing crisis. However, it is realistic to suggest that the establishment of the NHS undermined the government's ability to establish an industrial health service in another respect. The estimated cost of the NHS had been projected to fall between £108 million (the figure given in the 1944 White Paper) and £134 million (the calculation in the financial memoranda which was appended to the National Health Service Bills).[64] In the first nine months of the NHS, the cost of the NHS was calculated at £279.3 million, rising to £433 million for the financial year 1949–1950 and increasingly steadily throughout the 1950s, ahead of revised estimates and despite repeated endeavours made by the Labour government and subsequent Conservative governments to rein in costs. By 1956–1957, total expenditure on the NHS had reached £633 million. Faced with these spiralling costs, government committees considered axing services established within the NHS, such as the dental and ophthalmic services.[65] This was hardly a propitious climate in which to extend the NHS to encompass industrial medical services; little wonder then that the government looked to employers to take the initiative.

For many years it had been assumed that specific aspects of the factory environment necessitated the provision of health care services which were not required by employees in other working environments. A number of characteristics could be listed to support this contention, most notably the risk posed to health by working with potentially hazardous substances; the risk posed by working in poor conditions which would undermine health more subtly leading to greater susceptibility to viruses, fatigue, rheumatism, varicose veins, poor posture; the threat posed by physical hazards and the need for prompt first aid and emergency treatment of resulting accidents. Other factors which could be enumerated were the sheer size of many factories, the simple fact that people spent most of their waking life within such premises and the belief that employers should take responsibility for the health of workers while they were on factory premises. However, the growing awareness that factories were characterised by their diversity, combined with the renewed focus placed upon conditions of work in non-industrial work-places brought about by the establishment of the Gowers Committee, began to blur many of these boundaries. What did distinguish factories from other working environments? By the 1950s it was widely recognised that most units of industrial production were small-scale. Moreover, the application of industrial technology to other employment sectors led to a replication of 'industrial hazards' in other workplaces. For those seeking the establishment of an industrial health service, these developments complicated matters further. Now participants in the debate were considering not only whether the objective should be the creation of an industrial health, medical or hygiene service but also whether a comprehensive occupational health, medical or hygiene service should be established. While this prospect may have offered more comprehensive coverage, it also offered those who opposed the creation of national industrial health services an opportunity for further discussion and further delay.

Organisations frequently spoke of 'the Industrial Health Service' and 'the Industrial Medical Service'. When presenting an earlier version of this chapter I was pertinently reminded by one retired professor of industrial medicine that 'the' Industrial Health Service had never existed.[66] The same point was made by Charles Sisson, a civil servant within the Ministry of Labour and one of the joint secretaries of the Committee. When people think of a 'health service', he argued, they think of an integrated health service with a primarily medical character like the NHS. 'An Industrial Health Service in this sense', he wrote, 'does not exist. What does exist is a number of industrial health (including medical) services, statutory and non-statutory, under the control of this and other Departments and of private employers.'[67] In much the same way that the rhetoric of health and the idealised vision of the healthy factory had helped disguise the disparity in working conditions in the interwar years, the use of this phrase evoked a misleading picture of a nationwide service and could help undermine efforts to establish just such a service. Ultimately, the ideal of the healthy factory which had flourished in the interwar years was excluded and subsequently marginalised from national health services. The exclusion of industrial medical services had been justified on the grounds that they belonged to the sphere of industrial organisation, not personal health care. This supposition was vigorously challenged by the TUC and BMA which argued that the elevated risk to health posed by the combination of poor working conditions, hazardous processes and toxic materials necessitated the establishment of nationwide medical supervision within factories.

The findings of the 1972 Safety and Health at Work Committee, chaired by Lord Robens,[68] reflected how this earlier vision of the health factory had been lost and, in its place, a more circumscribed discourse of risk minimisation had triumphed. This Report argued that self-regulation was a more effective means to protect and raise standards of safety and health preservation at work than statutory regulations. It also exemplified how the fate of industrial health services was only considered in relation to the NHS. While acknowledging that health provisions in other countries were often effectively centred around the workplace, the Report concluded that there were insufficient resources both to do this and to retain the current system built up around private individuals. Its assessment of the relationship between occupational medicine and the NHS, and of the role of medical practitioners within the workplace, demonstrated that a narrow model of occupational hygiene had come to predominate. The factory was discarded as a site to forge health: it became relegated as an area which needed to be policed to prevent harm. The role of occupational medicine, this Report claimed,

> Can be understood only against the background of the general structure of health services in this country. The main element is the personal and mainly curative work of the NHS which is centred on the individual and family and not on the place of work... In the field of occupational health the working environment is of predominant importance, and it is engineers, chemists and others rather than doctors who have the expertise to change it.[69]

Conclusion

Industrial health; industrial illness

This study has attempted to complement the growing body of historical literature on occupational disease and death by studying health rather than disability. It has traced how a holistic concept of health which stressed the interaction of mind and body, prevention of disease and promotion of enhanced mental and physical vigour underpinned the approach adopted by the Health of Munition Workers Committee in the First World War. In the interwar years, the factory lay at the heart of a broadly based health movement. Traditionally viewed as a site that produced illness, it was now envisaged that the factory could be radically transformed into an attractive environment, able to create and promote health by providing workers with fresh air and sunlight, healthy meals and opportunities for sporting and recreational activities.

The moment when a state of health transforms into a state of illness is impossible to identify, an observation which links the history of industrial health (construed in its positive sense) to the history of industrial illness; after all, the term 'industrial health' tends to be used to describe the detection, treatment and prevention of disease caused by working materials, processes or environments. While the absence of disease is certainly a precondition of health, it is an impoverished definition if construed solely in these terms; certainly advocates of health promotion in the interwar years had something more ambitious and far-reaching in mind. The recent review of the health of the working population demonstrated how 'health' can be applied ambiguously when it refers to the desirability of enabling people with 'health conditions' to participate in the workforce, using the phrase 'health conditions' to denote illness or disability.[1]

207

The promotion of physical and mental health was a difficult objective to pursue because health was difficult to objectively see and quantify. Historians who have studied the history of industrial illness note that greater difficulty was experienced securing compensation for industrial illness than for workplace accidents. The 1897 Workmen's Compensation Act enabled employees who had become incapacitated through a workplace accident to secure compensation, though as P. W. J. Bartrip and S. B. Burman noted, 'the terms "arising out of and in the course of employment" and "on, or in, or about the works" were fruitful sources of litigation'.[2] Thus, even workers whose health was impaired by an accident struggled to demonstrate that their mishap was directly caused by their work and in the workplace, a debate that would persist in the interwar years due to the indistinct parameters of home and workplace. Moreover, as Bartrip and Burnam discussed, the failure of the 1897 Act to define what constituted an 'accident' provoked further litigation. A 1905 judgement went in favour of a workman who had contracted anthrax after the judge concluded that the introduction of anthrax into the body in the course of work constituted an industrial accident; subsequent industrial disease cases were dismissed on the grounds that the aetiology of the illness was too imprecise to enable it to be attributed, as an accident could be, to a specific time, place and event.[3]

The cause, onset and 'industrial' nature of a disease were more contentious issues. Bartrip and Burman noted that difficulties were experienced securing compensation in cases where an injury might have arisen out of a pre-existing injury or illness;[4] similarly, Arthur McIvor and Ronald Johnston examined the relative ease with which coal workers' pneumoconiosis was recognised as an industrial disease when contrasted with the obstacles facing those who attempted to secure compensation for miners incapacitated by bronchitis and emphysema.[5] The presence of the latter two conditions among the population at large introduced enough ambiguity to contest that responsibility for the illness lay in the management of the workplace.

'When one hears about another person's physical pain', wrote Elaine Scarry, 'the events happening within the interior of that person's body may seem to have the remote character of some deep subterranean fact, belonging to an invisible geography that, however, portentous, has no reality because it has not yet manifested itself on the visible surface of the earth.'[6] Scarry argued that the sufferer is isolated further by the inadequacy of language to express and communicate the nature of their pain to other people. She contends that our interior states of consciousness are linked to objects outside the body, indicating people's capacity 'to move out beyond the boundaries of his or her own body into the external, sharable world'. Pain is the exception to this rule as it has 'no referential content. It is not *of* or *for* anything. It is precisely because it takes no object that it, more than any other phenomenon, resists objectification in language.'[7] Given the invisibility of pain and the inadequacy of language to effectively communicate the nature of pain to other people, incapacity and sickness were ranked according to visibility and their explicit link to workplace conditions. Consequently it is unsurprising that workplace accidents, occurring in a definable location and time and usually inflicting visible physical damage, were ascribed greatest legitimacy. Conversely, disease developed insidiously and often invisibly, making it difficult to define the causative agent of sickness or the exact moment when one became ill: in this instance, greater legitimacy was ascribed to those illnesses only found within workers of a particular industry. Greater ambivalence surrounded health conditions which could be exacerbated by workplace conditions but which were also to be found in the population at large. Mental distress in response to workplace conditions was arguably the least visible, tangible or quantifiable.

The intangibility of health made it difficult to secure support and funding as it offered no immediate, visible or measurable return. Consequently there was a greater willingness to support efforts to remedy visible physical damage caused by accident or disease than to invest in preventive health promotion schemes or mental health services. This rationale undoubtedly explains trade union enthusiasm for medical services for foundries, a hazardous industry where tangible results could be demonstrated. Similarly, the willingness of trade unions to fund the Manor House Hospital and convalescent home provision for injured members testifies to the attractiveness of being seen to intervene on a definable, visible health problem.

Health could only be maintained and promoted if illness was prevented. As Thomas Legge observed, it was pointless to discuss methods workers could adopt to promote their own health through diet and exercise when, by and large, it was their working environment and working materials, factors which were controlled by their employers, which posed the most obvious threat to their health. Consequently, *The Rise and Fall of the Healthy Factory* has sought to elucidate attitudes towards the relationship between the health of the individual worker and the workplace environment. In her study of responses to the occupational health risks experienced by workers in the Lancashire cotton industry between 1870 and 1918, Janet Greenlees argued that the main obstacle blocking the reform of unhealthy working practices was the belief that the issue was one of personal responsibility; consequently public health interventions did not impinge on the workplace environment.[8] A shift in opinion can be detected in the reports of lady factory inspectors prior to the outbreak of the First World War, which voiced the opinion that a clean and hygienic workplace was a prerequisite if workers' health was to be maintained. Subsequently, the Health of Munition Workers Committee stressed the importance of establishing working conditions designed to optimise the physiological health and productivity of workers.

In the First World War and interwar years, the individualised health needs of workers became an important topic, increasingly conceptualised psychologically rather than physiologically. Describing the development of large-scale production methods, trade unionists and industrial welfare supervisors depicted the plight of the individual worker who had become simply a 'cog in the machine'; whose work was devoid of personal fulfilment. Could the factory and work processes be transformed to fulfil the physical and mental lives of workers at a time when alienating repetitive tasks displaced skilled craftsmanship? The question was one of adaptation; whether the working environment could be adapted to the individualised needs of the worker, if work could be fitted to his or her temperament and specially tailored medical and welfare services and mental distractions be provided, or whether individuals should be expected to adapt to their jobs and workplace. This study indicates that, lacking the resources to genuinely individualise the worker, industrial welfare sought to humanise industry by advocating the model of the healthy factory; in practice it stifled individualism by classifying individuals into groups and urging them to 'play for the team'. In the aftermath of the Second World War, the focus on the individual disappeared as the Factory Department and Ministry of Labour promoted a model of industrial hygiene. Once more, the health needs of the individual were subordinated to the health of the working environment.

Finally, it is worth noting that health promotion did not necessarily equate to medical intervention: if the heyday of the 'healthy factory' is examined in detail, we find that while innovative décor schemes, gardens and canteens may have flourished, only a tiny handful of factories had inaugurated any medical facilities beyond the provision of a first aid box, which in any case had become a legal requirement in all factories employing more than 50 employees under the 1924 Workmen's Compensation Act. This observation provides some support for the argument that the ideal of the healthy factory, with its emphasis on the psychological well-being of the worker, trivialised and marginalised the risk of industrial illness.

Trade unions and the politics of industrial health

A key objective of this monograph has been to provide an account of national politics underpinning developments in industrial health from the perspective of the Trades Union Congress (TUC), the national body which aimed to represent and advance the interests of workers throughout England and Wales.[9] This approach was adopted partly because the fate of industrial health services impinged most directly upon the health and livelihood of workers and partly because government records on industrial health services demonstrate that much of the impetus for the development of preventive occupational healthcare emanated from the TUC. It is not without its pitfalls, however; trade union density rarely reached even half the total workforce in the period under study, undermining the capacity of the TUC and its affiliated unions to speak for the entire workforce: Trade union density stood at 23 per cent of the workforce in 1914 at the starting point of this study, rising to 45.2 per cent in 1920, only to sink back to 25.4 per cent by 1930. Membership had recovered to 38.6 per cent by 1945 after which it remained just over the 40 per cent mark throughout the 1960s.[10] Studies of workplace practice, moreover, have demonstrated a marked divergence from official union advice on health and safety issues; even if trade unions chose to prioritise health and safety, workers could still reject precautions which impeded productivity and earnings or threatened to undermine a masculine workplace culture.[11] This study has also highlighted the tensions which arose when the long-term policy goals of the TUC clashed with the immediate, short-term solutions which local branches sought to implement.

While acknowledging these reservations, the archives of trade unions and the TUC still offer historians the most effective means to trace the efforts of workers to shape policy at a national level. Using these records to study the role played by the TUC and its affiliated unions in the promotion of workers' health and efforts to humanise the factory, in the context of an unfavourable economic climate and in an era when professional knowledge was given precedence, this study finds no basis for the assertion that wages and financial compensation were prioritised by unions at the expense of workers' health,[12] adding weight to a revisionist historiography which has begun to explore the proactive role played by trade unions in the prevention of industrial illness.[13] Joseph

Melling has suggested that the historical debate should advance beyond the question of whether trade unions pursued monetary compensation at the expense of safety, arguing that there was 'no intrinsic or necessary conflict between the pursuit of higher compensation rewards and the promotion of safety at work'.[14] Indeed, given the interrelationship of industrial health and illness discussed above, I feel that a more fruitful way to assess trade union responses is to see the pursuit of compensation and the pursuit of health as complementary, not competing. It is not the case that the TUC prioritised compensation payouts over the pursuit of health: the analysis of TUC work undertaken for this study revealed that it was one of the most enthusiastic advocates of industrial health services and adopted a variety of approaches in its attempts to enact its health promotion objectives in tandem with and as a natural adjunct to its work to extend the scheduling of industrial disease.[15] Indeed, as Chapter 5 demonstrated, the tenacity of TUC campaigning for a national occupational health service very nearly propelled a Conservative government to inaugurate such a service in 1954.

The TUC was not the only organisation seeking to promote its interests through industrial health policy and this monograph has considered the involvement of the British Employers' Confederation (BEC), medical associations, voluntary organisations and the State in the contestation and shaping of policy. It had indeed been hoped to incorporate more material on the role played by the BEC in the politics of industrial health; however, in contrast to the TUC archives, little material could be identified which demonstrated an active interest in positive health promotion among employers.[16] The correspondence files relating to the Industrial Health Services Committee (IHSC) did reveal reluctance among those employers who had established industrial medical services to relinquish these provisions;[17] in contrast, the files held by the BEC relating to the work of the Industrial Health Education Society demonstrated the BEC's disengagement with voluntary health promotion.[18] These findings largely resonate with conclusions arrived at by Mark Bufton and Joseph Melling, who argued that employers 'frequently attributed workers' illnesses to heredity, domestic or personal lifestyles, and infections or disabilities caused in the wider physical and social environment rather than the conditions found in industry'.[19] It was not inconsistent indeed for an employer who had inaugurated a welfare scheme within the factory to turn the tables and argue that the factory environment improved the health of their employees who irresponsibly jeopardised their health and consequently workplace productivity in their own time and own homes.

The era under consideration has been characterised by Keith Middlemas's influential study as a period of consensus politics, in which the State circumvented social unrest and political strife by incorporating employer and trade union bodies into governance.[20] Support for this line of argument could be drawn from *The Rise and Fall of the Healthy Factory*, which has demonstrated how developments in policy and practice emerged through the considerations of tripartite committees rather than the imposition of new legislation. Yet the picture which emerges is not one of harmonious discussion; frequently, it was disharmony and dissension rather than harmony and consensus which characterised the perspectives of the organisations involved.

The case study of industrial health indicates the constraints of developing policy through a model of consensus politics: the respective power held by different interested organisations was unequal, while the sheer number of committees established during the Second World War and post-war period, all with overlapping remits and memberships, tended to hinder rather than promote progress in the development of industrial health services. A frequent supposition expressed by voluntary bodies, medical organisations and the State was that if employers were made aware of the savings which could be made by sponsoring industrial health, most would do so without the need to exert any compulsion; by and large, this supposition proved ungrounded.

Given the lack of consensus displayed by employers, trade unions and medical organisations, the decision of successive governments to leave developments in the field of industrial health to voluntary enterprise appears woefully remiss. By shifting the focus from the State to the TUC, this monograph has sought to problematise the role played by the State and investigate why it did not adopt a more central and definitive role in the development of industrial health services. State involvement in industrial health was directly linked to the challenges experienced by the State in wartime and in particular the urgent necessity to maintain levels of industrial production if war was to be effectively prosecuted. During these times of crises, national health and industrial health, and by extension the health of the workforce, were perceived to be inexorably linked, providing 'ammunition' for David Edgerton's argument that Britain could be characterised as much as a Warfare State as a Welfare State in the era 1920 to 1970.[21] The shortage of labour in both World Wars made the protection of human capital an expedient concern for the State. Once the uncharacteristic period of crises posed by war had receded, the State's interest in industrial health likewise ebbed away.

The State may have been quick to jettison industrial health once peace had been restored; the ties between efficiency, prosperity and the health of the nation remained pervasive however. The trend of advertising healthy factories in which commodities were produced, discussed in Chapter 2, indicates how entwined consumption and production had become. The health and prosperity of the nation and the health of the nation's workers; the needs and interests of consumers and producers were perceived to be intimately bound up with one another. The Shredded Wheat advertisement, which featured the new model factory at Welwyn Garden City, typified this way of thinking, linking together healthy consumption and healthy production to the health of the nation.

Rethinking the factory in twentieth-century Britain

Industrial health services were designed to compensate for the hazards inherent in the factory environment; consequently, one would expect those organisations which participated in the debate on the future of industrial health services to have a thorough knowledge of the form and function of different factory environments. What has transpired in this study is that such a knowledge was very often absent and, moreover, that this neglect has largely been shared by historians.[22]

Reliance on a utilitarian definition of a factory as a site in which goods are produced utilising mechanisation misleadingly conceals the radical transformation of the factory in the course of the twentieth century. Factories remained, first and foremost, sites of production. They were also, however, permeable institutions, flexible to the prevailing social, cultural and economic trends which led to a reinterpretation of their form and function. Robert Owen's experiments indicate that the form and function of the factory began to be questioned at the very dawn of the system of factory production.[23] By the 1920s, enthusiasm and excitement characterised descriptions of new factories as industrial welfare advocates, trade unions, medical professionals and health promotion groups visualised a new world of utopian industrial production which would harmonise the interests of the worker, the manager and the environment. It was now no longer absurd to envisage the factory as a site of health production; instead a reciprocally beneficial circuit was envisaged whereby a well-designed factory would produce an enhanced state of health within its employees who would consequently find their own productive capacities had increased. The factory did not have to be a grim, dank building situated in an urban sprawl which sapped the health of its workers. Instead, it could be a place of beauty, a stimulating environment in which work could be humanised and monotony countered by invigorating the workers' senses. No longer did the factory have to be hidden away; now sleek, modernist crystal palaces, integrated into new suburban developments and garden cities, could be actively used to advertise the firm.

The very permeability of the factory and new beliefs as to its functions challenged preconceived ideas of what constituted its legitimate parameters and boundaries: the notion of the factory was in flux. Spatially, the factory was reconfigured to incorporate new rooms suited to its newly expanded range of functions. The introduction of canteens, washrooms, recreation spaces and rest rooms replicated the model of the worker's home: kitchen, bathroom, parlour and bedroom. Externally, the creation of new gardens blurred the boundaries of the factory. In part, these initiatives were introduced to meet the perceived needs of particular groups working within industry, particularly young people and women. These developments, intended to enhance and promote physical and mental well-being, introduced a certain ambiguity to the status of workers, exacerbated by the extension of welfare supervision into the homes of workers and the establishment of work transportation. Where and when did work start and end? At what stage did one become an employee and at what stage was one a private citizen? The establishment of works canteens created an incursion into workers' lunch hours; workmen's compensation cases were brought in which a worker received an injury while taking transportation provided by his employer to his workplace; while walking through work grounds on route to work; while undertaking errands off the factory site or while in the lunch hour.[24] The very permeability of the boundary between factory and home made it difficult to define what was an 'industrial' illness and what was simply an 'illness', as discussed above. Trade unions, employers and industrial health and welfare practitioners fiercely contested whether the work or the home environment produced sickness; it was frequently asserted that the sicknesses experienced by factory workers were best understood simply as ordinary illness which had manifested in an industrial environment. At stake in this debate was the question of which party should assume responsibility for illness and its treatment; and ergo, which should pay to provide such services.

Despite the endeavours of factory managers to foster company loyalty and a communal spirit with the inception of welfare facilities, the factory in the twentieth century remained a site of industrial relations.

Health services, based within factories and provided by managers for workers, were consequently tarnished, perceived as a tool of industrial relations. Certainly, there is a legitimacy to the line of argument that employers' decisions to implement welfare schemes could have been influenced by a desire to enhance worker loyalty and managerial control; as the TUC's 1932 survey indicated, the provision of financial insurance schemes held particular appeal to workers in a time of economic insecurity.[25] Indeed, this study has demonstrated that in the interwar years, when the economic downturn constrained what unions could offer to workers, that the pursuit of health services could help legitimate pre-existing industrial relations issues, refashioning the same matters as health issues. This is not to dismiss trade union endeavours in the field of health as a cynical ploy to bolster the authority of the TUC; the case study of the Industrial Health Education Society, a group which sought to separate health and politics, revealed just how enmeshed health in the workplace was with industrial politics.

The ideal of the healthy factory, the changing conceptions of what the form and function of a factory should be and the publicity given to the inauguration of new model, modernist factories disguised the diverse nature of premises which constituted a factory. Despite the move towards larger units of production, the small factory continued to predominate and while the institution of a medical service may have been both desirable and practicable at a factory employing 5,000 people, it was clearly not an option for a small works employing 25 staff members. It was only really during the Second World War that the problem of the small factory was identified. The development of grouped services appeared to offer a workable model to protect the health of workers engaged in small enterprises, but while the schemes established with the aid of Nuffield funding proved remarkably resilient, neither employers nor the government demonstrated the initiative to invest further in this approach.

The conflicting evidence given to the IHSC,[26] discussed in Chapter 6, indicates that employers and the Ministry of Labour still viewed industrial health services primarily as an intrinsic part of industrial relations rather than as a health service per se. On these grounds, it would prove difficult to incorporate industrial medical services into the NHS, a flagship product of consensus politics staffed by government employees that aimed to provide an objective, neutral service to all citizens. In contrast, the industrial health services, staffed by personnel employed by the management, were viewed as a subjective tool of industrial relations. No one questioned that the NHS provided a health care service to citizens which should legitimately be operated by the Ministry of Health; once the citizen entered the industrial workplace, however, they were seemingly divested of their citizenship. Reconceived of as a cog in a machine, whose impairment threatened the smooth functioning of the entire enterprise, health services were viewed through this prism as an industrial matter, not a health matter, best left in the hands of private enterprise as an industrial relations issue to be overseen, but not controlled, by the Ministry of Labour.

The fall of the healthy factory; the fall of the factory?

This monograph has traced the rise and fall of the healthy factory through to 1960, by which stage it was apparent that a comprehensive occupational health service as envisaged by the IHSC was unlikely to be implemented by the government. In 1972, the work of medical factory inspectors, appointed factory doctors and the medical advisers of employment rehabilitation centres were duly reconfigured into the Employment Medical Advisory Service, a new national advisory resource intended to identify health hazards within industry. The Society of Occupational Medicine, which still urged the establishment of a service within the Department of Health from which a national occupational health service could evolve, argued that the narrow remit of this new Service threatened to isolate industrial medicine further.[27] In 1974, the Employment Medical Advisory Service was absorbed into the newly established Health and Safety Executive.

The 1972 Report of the Health and Safety at Work Committee argued that self-regulation and responsibility would be a more effective means to protect and raise standards of safety and health preservation at work than statutory regulations.[28] While acknowledging that the health services of other countries were centred round the workplace, the Report authors concluded that there were insufficient resources to do both this and retain the current NHS system built up around private individuals, envisaging instead a narrow remit for the occupational health services. In so doing, it cemented the boundary between personal health care services and occupational health which had been created with the establishment of the NHS and consigned the ideal of the healthy factory to history. The resulting 1974 Health and Safety at Work Act discarded the detailed regulations which had characterised the 1937 Factory Act in favour of generalised rules which had typified the 1901 Factory Act, on the grounds that specific regulations rapidly became obsolete. The 1974 Act outlined responsibilities to be undertaken by employers 'so far as is reasonably practicable', introducing a certain subjectivity and ambiguity; similarly, the 1901 Act had been peppered with the phrases 'sufficient and suitable', 'reasonable' and 'adequate means'. The 1901 Act had hindered the ability of factory inspectors to secure convictions for conditions which they considered hazardous because of the leeway contained within the phraseology of the legislation;[29] the 1974 Act, which largely relied on employers and workers to regulate workplace conditions, was feebly enforced by an understaffed Factory Inspectorate with minimal powers.[30]

As hopes for a comprehensive industrial health service receded, the fortunes of British industry also declined. Britain's share of world exports of manufactures fell from 24.6 per cent in 1950 to 9.1 per cent in 1973.[31] Manufacturing as a percentage share of annual gross domestic product fell from 32.1 per cent in 1960 to 24.9 per cent in 1979 and 18 per cent by 1989; conversely, the services sector expanded from 53.6 per cent in 1960 to 69.4 per cent by 1989.[32] The provision of a comprehensive health care service was seen as neither economically viable nor a priority within a decaying industrial sector which struggled to remain competitive in a global market.

Given the insidious onset of many industrial diseases, the fatal legacy arising from the neglect of serious health risks and marginalisation of occupational health following the inauguration of the NHS continues to unfold.[33] However, the diseases which had emerged in the industrial workplace, associated with toxic materials and hazardous working environments, are slowly being displaced by new occupational health problems. In her recent study of contemporary British work culture, Madeleine Bunting noted that the historic decline in working hours had gone into reverse in the past two decades.[34] An ethos of efficiency prompted job intensification, increasing workloads and a corresponding rise in exhaustion and work-related stress. Bunting found that for one in three British workers, exhaustion or stress had become an inescapable part of their working lives;[35] work-related stress, depression and anxiety had become the main cause of absence from work on health grounds, accounting for 13.4 million lost working days per year.[36] These health problems plagued white collar workers, who found that the boundaries between work and private time had been eroded by new technology and communication methods. Increasingly, British workers were not fully utilising their annual holiday leave, a right British trade unions had fought for in the interwar years; rest breaks and lunch breaks within the working day were also shrinking. Bunting argued that companies had been able to extract greater loyalty and commitment from their employees by fostering a sense of brand loyalty, urging employees to see work as a vocation which aligned personal interests and the work ethic. Measured by economic standards, the result of the emotion and time poured into work has however been disappointing; noting that productivity levels in Britain fell beneath levels in America and Europe, where shorter hours were worked, Bunting urged the government to reverse its stance of minimal regulation and enforce shorter working hours.

While the British workforce records unprecedented levels of work-related stress, the illness burden of industrial production has not so much been resolved as relocated. The manufacturing of products for the British market has, in many instances, been transferred to countries where production costs are lower. In some cases, products sold in Britain are made in conditions which resemble more closely the British factory of the late nineteenth rather than the late twentieth century. In June 2008, child labour was found to be used in the manufacture of clothing for the British firm Primark;[37] the most recent example of a major firm whose products were manufactured in conditions which did not meet their own code of practice.[38] Similarly, Chris Sellers has recently highlighted how the circulation and consequent health burden of lead had been globally redistributed, discussing the discovery of lead paint in children's toys for sale in America, an incident which sparked outrage not at the health risks posed to the producers but the risks posed to American consumers.[39] These recent examples indicate that consumption and production patterns remain intimately bound together; the demand for cheap goods fuelling low cost, hazardous production methods. If the exclusion of industrial health services from the NHS perpetuated work-based health inequalities, as McIvor and Johnston suggest,[40] the inequalities of health arising from consumption and production have now been redistributed on a global scale.

The re-emergence of the healthy workplace? Dame Carole Black's Review of the Health of the Working Age Population[41]

This book has presented a story of the rise and fall of the ideal of the healthy factory. However, recent government health policy has been more in keeping with the ideals of health promotion which characterised the interwar years. Lord Darzi's recent report on the future of the NHS, for example, was couched in the language of health promotion, preventive care and well-being.[42] Identifying six key areas which he felt were imperative to address if health and well-being were to be enhanced – obesity, alcohol harm, drug addiction, smoking rates and sexual and mental health – Darzi stressed the role of lifestyle choices and personal behaviour, once more placing the responsibility of the individual for their own health centre stage.[43]

In keeping with this trend, the publication of Dame Carole Black's Review in March 2008, with which this monograph opened, indicates that the workplace is once more being viewed as a potential site of health production. Delineating the scale of the problem, Black drew on quantitative measures of expense: a traditional method used to counter the intangibility of health by demonstrating the costs of failing to preserve it. The cost to tax payers, including benefit costs, medical expenses and lost taxes, was estimated at over £60 billion, while the annual economic cost of sickness absence and worklessness associated with working age ill-health was put at over £100 billion. Black noted that this figure was equivalent to the GDP of Portugal,[44] a choice of comparison which indicates that the value of health services for workers is still calculated in relation to levels of economic productivity. The Review implies that much of the initiative will be left in the hands of businesses, as Black believed that employers would be willing to implement health schemes if 'a robust model for measuring and reporting on the benefits of employer investments in health and well-being' was established.[45] Black also envisaged a role for trade unions, which she argued, 'must seize the opportunity to champion health and well-being in the workplace'.[46]

Black's Review was in line with recent government policy on employment, welfare and social exclusion; in short, Black believed that it was possible to align social justice and economic goals by improving the health of the working population.[47] Consequently, her Review explored both the health problems which prevented people entering employment as well as those which caused people to leave the workforce. Black's stance was informed by her belief that work could be good for health, able to reverse the harmful effects of long-term unemployment and prolonged sickness absence. Black noted that each year 600,000 people move onto incapacity benefits; 200,000 of which have a mental health condition (this latter figure had not changed over the past decade).[48] She suggested that attitudes might be transformed if the traditional sick note were replaced with a 'fit note' which would list what people were able to do rather than what they were incapable of.[49] Black also advocated the establishment of a Fit for Work service which she envisaged would deliver higher tax receipts, reduced benefit payment, better workplace productivity and, over time, 'reduced costs to the NHS'.[50]

This is a tacit admission that the exclusion of industrial health services from the NHS and the subsequent neglect of occupational health issues by the government have had steep costs to industry, to the nation and to individual health and well-being.

Black's Review reflected on the legacy which resulted from the exclusion of occupational health from the NHS, acknowledging that the marginalisation of occupational health led to its decline, characterised by a diminishing workforce, a shrinking academic base and a lack of quality data. Understanding of work-related illness among mainstream medicine suffered as a result. Black claimed that GPs frequently lack a sufficient understanding of the relationship between work and a patient's health to be able to give effective advice. 'With many employers to date having failed to provide access to adequate occupational health, and the associated costs to the taxpayer and the economy being so substantial', Black wrote, 'there is a strong case for the NHS being involved in the provision of these work-related health interventions.'[51]

Many of the ideals of the healthy factory are reflected in Black's Review, albeit in a different economic and political framework. Once more, the workplace has been construed as a site of health production and the employee conceptualised as an individual whose health is intimately related to their social environment within and outside the workplace. The health of the individual, the nation and industrial prosperity are once more in alignment. Beyond a shift in rhetoric from sick to fit and illness to health, one might question whether Black's expectation that employers will implement health services is realistic and ask whether the workplace can be adequately transformed into a site of health promotion without more extensive government intervention and an effective interlinking of NHS and occupational health services.

Notes

Introduction

1. C. Black, *Working for a Healthier Tomorrow: Review of the Health of Britain's Working Age Population* (London, 2008), p. 10. Black is the National Director for Health and Work and the Review was commissioned by the government.
2. Discussed in Chapter 3, 'The Role of the Workplace in Health and Well-Being', ibid., pp. 51–60.
3. These changes came into effect on 6 April 2010. See www.workingfor-health.gov.uk/Initiatives/Reforming-the-medical-statement/Default.aspx, consulted 29 April 2010.
4. For further details, see www.workingforhealth.gov.uk/Initiatives/fit-for-work-service/Default.aspx, consulted 29 April 2010.
5. Black, *Working for a Healthier Tomorrow*, p. 12. Details of government initiatives established in response to the Review can be found at www.workingforhealth.gov.uk/, consulted 29 April 2010.
6. C. U. Kerr, 'A New Model Factory', *Journal of Industrial Welfare*, 8 (1926), 160–164.
7. D. Proud, *Welfare Work: Employers' Experiments for Improving Working Conditions in Factories* with Foreword by David Lloyd George PM and first Minister of Munitions (3rd edition, London, 1918), p. 256.
8. Editorial, 'The Human Machine', *The Times*, 25 September 1920, 11.
9. A. Meiklejohn, 'Sixty Years of Industrial Medicine in Great Britain', *British Journal of Industrial Medicine*, 13 (1956), 155–162; 160–161.
10. D. Lloyd George, 'Foreword', in Proud, *Welfare Work*, p. ix.
11. *Report of a Committee of Enquiry on Industrial Health Services* (1951), Cmd 8170, p. 14.
12. *Safety and Health at Work* (1972), Cmnd 5034.
13. Though not all technological innovations occurred in the new factories while the new factories could retain aspects of traditional handicraft: cotton manufacturers in the eighteenth and early nineteenth centuries continued to employ domestic weavers. For example, see M. Berg, *The Age of Manufactures 1700–1820: Industry, Innovation and Work in Britain* (London, 1994).
14. T. G. Geraghty, 'Factory System', in J. Mokyr (ed.), *Oxford Encyclopaedia of Economic History Vol. 2* (Oxford, 2003), pp. 247–253.
15. A. Ure, *The Philosophy of Manufactures: or, An Exposition of the Scientific, Moral, and Commercial Economy of the Factory System of Great Britain* (2nd edition, London, 1835), p. 13.
16. P. Joyce, *Work, Society and Politics: The Culture of the Factory in Later Victorian England* (London, 1982), p. 52.
17. W. Hamish Fraser, *A History of British Trade Unionism 1700–1998* (London, 1999), pp. 13–14.

18. See R. M. Martin, *TUC: The Growth of a Pressure Group 1868–1976* (Oxford, 1980) and D. F. MacDonald, *The State and the Trade Unions* (2nd edition, London, 1976).
19. J. F. C. Harrison, *Robert Owen and the Owenites in Britain and America: The Quest for the New Moral World* (2nd edition, Aldershot, 1994), p. 153.
20. S. Phillips, 'Industrial Welfare and Recreation at Boots Pure Drug Company 1883–1945' (PhD thesis, Nottingham Trent University, 2003).
21. J. Melling, 'Employers, Industrial Welfare, and the Struggle for Work-Place Control in British Industry, 1880–1920', in H. F. Gospel and C. R. Littler (eds), *Managerial Strategies and Industrial Relations: An Historical and Comparative Study* (London, 1983), pp. 55–81.
22. See A. J. McIvor, 'Employers, the Government, and Industrial Fatigue in Britain, 1890–1918', *British Journal of Industrial Medicine*, 44 (1987), 724–732.
23. For an overview, see E. Crooks, *The Factory Inspectors: A Legacy of the Industrial Revolution* (Stroud, 2005), pp. 36–56.
24. H. Martindale, *From One Generation to Another 1839–1944: A Book of Memoirs* (London, 1944), p. 143. Helen Jones has argued that due to the small number employed, women inspectors in the nineteenth and early twentieth centuries were more successful in advancing the role of middle-class women in the labour market than in assisting women workers: H. Jones, 'Women Health Workers: The Case of the First Women Factory Inspectors in Britain', *Social History of Medicine*, 1 (1988), 165–181.
25. See G. Darley, *Factory* (London, 2003).
26. In 1949, for example, 202,868 of the 243,769 factories in Britain employed fewer than 26 people: *Report of a Committee of Enquiry on Industrial Health Services*, Appendix E, p. 28.
27. The piecemeal nature of the existing legislation which had been established to counter problems as they emerged was acknowledged by the Robens Report: *Safety and Health at Work* , p. 5. Carolyn Malone and Barbara Harrison have examined the debate surrounding the hazards of women's labour which informed protective labour legislation that restricted or excluded women from certain trades. See C. Malone, *Women's Bodies and Dangerous Trades in England 1880–1914* (Woodbridge, 2003); B. Harrison, 'Suffer the Working Day – Women in the "Dangerous Trades," 1880–1914', *Women's Studies International Forum*, 13 (1990) 79–90 and B. Harrison, *Not only the 'Dangerous Trades': Women's Work and Health in Britain, 1880–1914* (Abingdon, 1996).
28. A. S. Wohl, 'The Canker of Industrial Disease', in *Endangering Lives: Public Health in Victorian Britain* (London, 1984), pp. 257–284.
29. S. Sturdy, 'The Industrial Body', in R. Cooter and J. Pickstone (eds), *Companion to Medicine in the Twentieth Century* (London, 2003), pp. 217–234.
30. R. Oastler, *The Factory Question: The Law or the Needle* (London, 1836), p. 1.
31. See R. Gray, *The Factory Question and Industrial England, 1830–1860* (Cambridge, 1996).
32. C. Turner Thackrah, *The Effects of the Principle Arts, Trades and Professions, and of Civic States and Habits of Living, on Health and Longevity* (London, 1831).
33. Thackrah, *The Effects of the Principle Arts*, p. 5.

34. R. Johnston and A. McIvor, *Lethal Work: A History of the Asbestos Tragedy in Scotland* (East Linton, 2000); G. Tweedale, *Magic Mineral to Killer Dust: Turner and Newall and the Asbestos Hazard* (Oxford, 2000); A. McIvor and R. Johnston, *Miners' Lung: A History of Dust Disease in British Coal Mining* (Aldershot, 2007).

35. P. W. J. Bartrip, *Workmen's Compensation in Twentieth Century Britain: Law, History and Social Policy* (Aldershot, 1987); P. W. J. Bartrip and S. B. Burman, *The Wounded Soldiers of Industry: Industrial Compensation Policy 1833–1897* (Oxford, 1983); M. W. Bufton and J. Melling, ' "A Mere Matter of Rock": Organised Labour, Scientific Evidence and British Government Schemes for Compensation of Silicosis and Pneumoconiosis among Coalminers, 1926–1940', *Medical History*, 49 (2005), 155–178.

36. J. C. Riley, *Sick, Not Dead: The Health of the British Workingmen During the Mortality Decline* (Baltimore, 1997).

37. C. Sellers, *Hazards of the Job: From Industrial Disease to Environmental Health Science* (Chapel Hill and London, 1997).

38. Melling, 'Employers, Industrial Welfare, and the Struggle for Work-Place Control'.

39. *Safety and Health at Work*, p. xv.

40. Sellers, *Hazards of the Job*, p. 8.

41. Sturdy, 'The Industrial Body', pp. 219–220.

42. J. Pickstone, 'Production, Community and Consumption: The Political Economy of Twentieth-Century Medicine', in Cooter and Pickstone (eds), *Companion to Medicine in the Twentieth Century*, pp. 1–19.

43. See, for example, J. Lewis, *The Politics of Motherhood: Child and Maternal Welfare in England 1900–1939* (London, 1980); D. Dwork, *War is Good for Babies and Other Young Children: A History of the Infant and Child Welfare Movement in England 1898–1918* (London, 1989); A. Davin, 'Imperialism and Motherhood', *History Workshop Journal*, 5 (1978), 9–66.

44. C. Gordon, *Dead on Arrival: The Politics of Health Care in Twentieth-Century America* (Princeton, 2005).

45. Pickstone, 'Production, Community and Consumption', p. 3.

46. This line of argument reflects the broader thesis advanced by Keith Middlemas, who argued that British governments after the First World War secured political harmony by incorporating major interest groups into the governing process. See K. Middlemas, *The Politics of Industrial Society: The Experience of the British System Since 1911* (London, 1979).

47. P. Weindling, 'Linking Self Help and Medical Science: The Social History of Occupational Health', in P. Weindling (ed.), *The Social History of Occupational Health* (London, 1985), pp. 2–31; p. 10.

48. Tweedale, *Magic Mineral to Killer Dust*, p. 289.

49. M. W. Bufton and J. Melling, 'Coming Up for Air: Experts, Employers, and Workers in Campaigns to Compensate Silicosis Sufferers in Britain, 1918–39', *Social History of Medicine*, 18 (2005), 63–86; 85.

50. McIvor and Johnston, *Miners' Lung*, pp. 185–233.

51. S. Bowden and G. Tweedale, 'Mondays without Dread: The Trade Unions Response to Byssinosis in the Lancashire Cotton Industry in the Twentieth Century', *Social History of Medicine*, 16 (2003), 79–96; 94–95; Johnston and McIvor, *Lethal Work*.

52. J. Melling, 'The Risks of Working and the Risks of Not Working: Trade Unions, Employers and Responses to the Risk of Occupational Illness in British Industry, c. 1890–1940s', *ESRC Centre for Analysis of Risk and Regulation Discussion Paper 12* (2003), pp. 14–34.

53. J. Stevenson, 'Occupations, Work and Organised Labour', in Stevenson, *The Penguin Social History of Britain: British Society 1914–45* (London, 1984), pp. 182–202; R. Price and G. S. Bain, 'The Labour Force' in A. H Halsey (ed.), *British Social Trends Since 1900* (Basingstoke, 2000), pp. 162–201.

54. H. Hendrick, *Images of Youth: Age, Class and the Male Youth Problem, 1880–1920* (Oxford, 1990).

55. See Davin, 'Imperialism and Motherhood'; Lewis, *The Politics of Motherhood*.

56. V. Long and H. Marland, 'From Danger and Motherhood to Health and Beauty: Health Advice for the Factory Girl in Early Twentieth-Century Britain', *Twentieth Century British History*, 20 (2009), 454–481.

57. M. Thomson, *Psychological Subjects: Identity, Culture and Health in Twentieth-Century Britain* (Oxford, 2006), pp. 142–150.

58. See R. McKibbin, 'The "Social Psychology" of Unemployment in Inter-War Britain', in *The Ideologies of Class: Social Relations in Britain 1880–1950* (London, 1991), pp. 228–258; N. Whiteside, 'Counting the Cost: Sickness and Disability among Working People in an Era of Industrial Recession, 1920–39', *Economic History Review*, 40 (1987), 228–246.

59. Wohl, 'The Canker of Industrial Disease'.

60. B. Harris, 'Educational Reform, Citizenship and the Origins of the School Medical Service', in M. Gijswijt-Hofstra and H. Marland (eds), *Cultures of Child Health in Britain and the Netherlands in the Twentieth Century* (Amsterdam and Atlanta, 2003), pp. 85–102.

61. See D. Porter, *Health, Civilisation and the State: A History of Public Health from Ancient to Modern Times* (London, 1999), pp. 111–162.

62. G. Newman, *The Rise of Preventive Medicine* (Oxford, 1932), p. 228. Italics in original text.

63. *Health of Munition Workers Committee Final Report: Industrial Health and Efficiency* (1918), Cd 9065, p. 15.

64. S. Murphy, 'The Early Days of the MRC Social Medicine Research Unit', *Social History of Medicine*, 12 (1999), 389–406; J. Lewis and B. Brookes, 'The Peckham Health Centre, "PEP", and the Concept of General Practice during the 1930s and 1940s', *Medical History*, 27 (1983), 151–161.

65. Murphy argues that 'support for social medicine was fuelled by the general increase in interest in social issues as well as the more specific wish to promote occupational health research': 'The Early Days', p. 394.

1 War and Industrial Health: The Productive Alliance

1. R. Squire, *Thirty Years in the Public Service: An Industrial Retrospect* (London, 1927), pp. 187–188.

2. See D. Fraser, *The Evolution of the British Welfare State* (2nd edition, London, 1992), p. xxi.

3. P. Thane, *Foundations of the Welfare State* (2nd edition, Harlow, 1996), p. xiii.

4. Fraser, *The Evolution of the British Welfare State*, p. 1.
5. J. Habermas, *The Structural Transformation of the Public Sphere: An Inquiry into a Category of Bourgeois Society* (trans. Thomas Burger, London, 1989); pp. 222–235 deals explicitly with the emergence of the Social Welfare State.
6. Thane, *Foundations of the Welfare State*.
7. G. Finlayson, *Citizen, State, and Social Welfare in Britain, 1830–1990* (Oxford, 1994), pp. 201–286.
8. S. Sturdy (ed.), *Medicine, Health and the Public Sphere in Britain 1600–2000* (London, 2002); N. Crossley and J. M. Roberts (eds), *After Habermas: New Perspectives on the Public Sphere* (Oxford, 2004); C. Calhoun (ed.), *Habermas and the Public Sphere* (Massachusetts, 1999). Crossley, in his study of resistance to psychiatry in the second half of the twentieth century, analysed the actions of multiple agents engaged in a critique of psychiatric practice through a model of social movement organisations: see N. Crossley, *Contesting Psychiatry: Social Movements in Mental Health* (London, 2006).
9. This chapter specifically explores the extent to which the state intervened to secure the health and welfare of factory workers; for an account of state intervention in the workplace more generally in both wars, see A. McIvor, *A History of Work in Britain, 1880–1950* (Hampshire, 2001), pp. 151–170.
10. C. W. Baker, *Government Control and Operation of Industry in Great Britain and the United States during the World War* (Oxford, 1921), p. 5.
11. K. Middlemas, *The Politics of Industrial Society: The Experience of the British System Since 1911* (London, 1980), pp. 276–277.
12. D. Edgerton, *Warfare State: Britain, 1920–1970* (Cambridge, 2006), p. 4; J. Pickstone, 'Production, Community and Consumption: The Political Economy of Twentieth-Century Medicine' in R. Cooter and J. Pickstone (eds), *Companion to Medicine in the Twentieth Century* (London, 2003), pp. 1–20.
13. *Annual Report of the Chief Inspector of Factories and Workshops for the Year 1914* (1915) Cd 8051, p. 39.
14. *Annual Report of the Chief Inspector of Factories and Workshops for the Year 1918* (1919), Cmd 340, pp. 31–33 describes these orders but does not provide detailed references. Welfare provisions including protective clothing, first aid arrangements, a cloakroom, mess room and washing facilities were detailed under the Tanning (Two-Bath Process) Welfare Order 1918 S.R. & O. 1918/368 and Dyeing (Use of Bichromate of Potassium or Sodium) Welfare Order 1918 S.R. & O. 1918/369.
15. *Health of Munition Workers Committee Interim Report: Industrial Efficiency and Fatigue* (1917), Cd 8511, p. 2.
16. *Annual Report of the Chief Inspector of Factories and Workshops for the Year 1918* (1919), Cmd 941, p. 33.
17. Editorial, 'Factory Work and Fatigue', *The Times*, 20 September 1915, 9.
18. The second son of chocolate manufacturer Joseph Rowntree, Seebohm worked within the family firm which was renowned for its industrial welfare. He produced sociological accounts of poverty and labour management, including S. Rowntree, *The Human Needs of Labour* (London, 1918). For an overview, See B. Harrison, 'Rowntree, (Benjamin) Seebohm (1871–1954)', *Oxford Dictionary of National Biography*, (Oxford, 2004). www.oxforddnb.com/view/article/35856, consulted on 22 February 2008.

19. *Annual Report of the Chief Inspector of Factories and Workshops for the Year 1918*, p. 36.
20. D. Proud, *Welfare Work: Employers' Experiments For Improving Working Conditions in Factories with Foreword by David Lloyd George PM and first Minister of Munitions* (3rd edition, London, 1918), p. 37.
21. *The Annual Report of the Chief Inspector of Factories and Workshops for the Year 1918*, p. 28.
22. E. Bevin, 'First Meeting of the Industrial Health Advisory Committee: Speech Delivered by the Rt Hon Ernest Bevin MP' (London, 1943), p. 5.
23. R. S. F. Schilling, 'Industrial Health Research: The Work of the Industrial Health Research Board', *British Journal of Industrial Medicine*, 1 (1944), 145–152.
24. Factories (Medical and Welfare Services) Order S.R. & O. 1940/1325.
25. Factories (Canteens) Order S.R. & O. 1940/1993.
26. These regulations encapsulated recommendations made by the Ministry of Labour and National Service's Departmental Committee on Lighting in Factories, *Lighting in Factories: Fifth Report* (London, 1940).
27. Select Committee on National Expenditure, *Third Report: Health and Welfare of Women in War Factories* (London, 1942), p. 3.
28. Modern Records Centre, University of Warwick, TUC Archive, MSS.292/141.2/5, questions and answers from the House of Commons, 25 March 1943.
29. *Annual Report of the Chief Inspector of Factories and Workshops for the Year 1919* (1920), Cd 941, p. 87.
30. Ibid, p. 112.
31. D. Thom with A. Ineson, 'TNT Poisoning and the Employment of Women Workers in the First World War', in D. Thom, *Nice Girls and Rude Girls: Women Workers in World War 1'* (2nd edition, London, 2000), pp. 122–143; p. 128.
32. In the interwar years, a small increase in the number of doctors and nurses working within industry can be documented; this is discussed in Chapter 2. Efforts to extend medical training on industrial health are explored in Chapter 3.
33. *Annual Report of the Chief Inspector of Factories for the Year 1942* (1943), Cmd 6471, p. 29.
34. *Annual Report of the Chief Inspector of Factories for the Year 1943* (1944), Cmd 6563, p. 53.
35. *Annual Report of the Chief Inspector of Factories for the Year 1944* (1945), Comd 6698, p. 53.
36. Ibid, p. 93. Of these, 179,939 were factories with electrical power, and 48,837 were factories without power, classified prior to 1938 as workshops.
37. *Annual Report of the Chief Inspector of Factories for the Year 1943*, p. 53.
38. 'Introductory', *British Journal of Industrial Medicine*, 1 (1944), 66.
39. *Annual Report of the Chief Inspector of Factories for the Year 1940* (1941), Cmd 6316, p. 23.
40. *Annual Report of the Chief Inspector of Factories for the Year 1942*, p. 28.
41. *Annual Report of the Chief Inspector of Factories for the Year 1943*, p. 21.
42. H. A. Waldron stated that the number of full-time doctors had declined to 51 by 1951, citing the Table B in *Report of a Committee of Enquiry on Industrial*

Health Services (1951), Cmd 8170, p. 29 as his source. This number however referred only to the number of doctors who held full-time posts as
Appointed Factory Doctors (i.e. the doctors appointed to undertake medical
examinations of young workers to meet the legal requirements of the 1937
Factory Act). The number of doctors employed voluntarily by employers on
a full-time basis as industrial medical officers was given in the same table
of the report as 186: H. A. Waldron 'Occupational Health during the Second
World War: Hope Deferred or Hope Abandoned?' *Medical History*, 41 (1997),
197–212; 204.

43. In 1943, the Factory Inspectorate noted that 8,395 nurses were employed
    in factories, of which around half were state registered: *Annual Report of
    the Chief Inspector of Factories for the Year 1943*, p. 53. The 1951 figures are
    derived from *Report of a Committee of Enquiry on Industrial Health Services*,
    p. 7. Of these 4,000, 2,600 were state registered nurses.
44. See B. Harrison, *Not only the 'Dangerous Trades': Women's Work and Health
    in Britain, 1880–1914* (Abingdon, 1996); C. Malone, *Women's Bodies and the
    Dangerous Trades in England, 1880–1914* (Woodbridge, 2003).
45. *Health of Munition Workers Committee Final Report: Industrial Health and
    Efficiency* (1918) Cd 9065, p. 7.
46. Ibid, p. 11.
47. Ibid, p. 15.
48. TUC Archive, MSS.292/141.2/5, E. Bevin, press release circulated by the
    Ministry of Labour and National Service, 11 March 1943.
49. 'The Doctor in Dungarees', *Public Health*, 56 (1943), 88.
50. TUC Archive, MSS.292/141.2/5, minutes of the first meeting of the IHAC,
    5 April 1943.
51. Statistics from the annual reports of the Factory Inspectorate for 1940,
    p. 24 and 1943, p. 52. Bridge's thoughts on dermatitis and washing were
    given in *The Annual Report of the Chief Inspector of Factories for the Year 1941*
    (1942), Cmd 6397, p. 22. Given the remit of this study, the impact of war-
    time conditions on the rates of industrial illness are largely excluded from
    study; a brief overview is provided in Waldron, 'Occupational Health dur-
    ing the Second World War', pp. 202–205, while an analysis of the impact
    of wartime conditions on occupational health and safety in Scotland is
    provided by R. Johnston and A. McIvor, 'The War and the Body at Work:
    Occupational Health and Safety in Scottish Industry, 1939–1945', *Journal of
    Scottish Historical Studies*, 24 (2004), 113–136.
52. *Annual Report of the Chief Inspector of Factories for the Year 1939* (1940), Cmd
    6251, p. 20.
53. TUC Archive, MSS.292/141.2/5, memoranda submitted to the second
    meeting of the Industrial Health Advisory Committee, 14 October 1943.
54. TUC Archive, MSS.292/141.2/5, 'Notes on "Rheumatism" in Industry', dis-
    cussed at the second meeting of the IHAC, 14 October 1943; this referred to
    the work of Dr Loughton Scott, who gave injections to dissolve the fibrous
    tissue which pressed on the nerve.
55. TUC Archive, MSS.292/141.2/5, notes on mass radiography, discussed at the
    second meeting of the IHAC, 14 October 1943.
56. TUC Archive, MSS.292/141.2/5, Ministry of Labour and National Service,
    Fighting Fit in the Factory (London, no date).

57. See S. MacDonald, *Simple Health Talks with Women War Workers* (London, 1917); B. Webb, *Health of Working Girls: A Handbook for Welfare Supervisors and Others* (London, 1917); D. J. Collier, *The Girl in Industry* (London, 1918).
58. TUC Archive, MSS.292/141.2/3, minutes of the Factory Sub-Committee of the Central Council for Health Education, 4 February 1943.
59. TUC Archive, MSS.292/141.2/4, R. Sutherland, 'Promoting Industrial Health: the Joint Responsibility of Management and Workpeople', *Labour Management* (August/September, 1944) reprint with no page numbers.
60. TUC Archive, MSS.292/141.2/3, minutes of the Factory Sub-Committee, 31 March 1943.
61. TUC Archive, MSS.292/844.1/5, memorandum of an interview between Mrs Hulme, Miss Evelyn (British Social Hygiene Council) and Mr Tewson, Mr Wray (TUC), 5 March 1940.
62. *Annual Report of the Chief Inspector of Factories and Workshops for the Year 1913* (1914), Cd 7491, p. 81.
63. *HMWC Final Report*, 'Section IX: Food and Canteens', pp. 51–61.
64. *HMWC Interim Report*, p. 104.
65. *The Annual Report of the Chief Inspector of Factories and Workshops for the Year 1914*, p. 52.
66. *Annual Report of the Chief Inspector of Factories for the Year 1940*, p. 14.
67. *Annual Report of the Chief Inspector of Factories for the Year 1943*, p. 60.
68. *Annual Report of the Chief Inspector of Factories for the Year 1944*, p. 84.
69. S.R. & O. 1943/573.
70. *Annual Report of the Chief Inspector of Factories for the Year 1943*, p. 64.
71. J. Vernon, *Hunger: A Modern History* (Cambridge and London, 2007), pp. 161–80l.
72. *Annual Report of the Chief Inspector of Factories for the Year 1943*, pp. 60–61; p. 61.
73. 'Section XIV: Cleanliness, Ventilation, Heating and Lighting', in *HMWC Final Report*, pp. 83–89.
74. *Lighting in Factories: Fifth Report*.
75. *Annual Report of the Chief Inspector of Factories and Workshops for the Year 1942*, p. 18.
76. *Annual Report of the Chief Inspector of Factories and Workshops for the Year 1943*, p. 31.
77. Mass Observation, *War Factory: A Report* (London, 1943), pp. 62–63.
78. *HMWC Final Report*, p. 15.
79. *HMWC Interim Report*, pp. 111.
80. Ibid, p. 12.
81. Ibid, p. 17.
82. Ibid, p. 32.
83. 'Girl Munition Workers. Health and a Twelve-Hours Day', *The Times*, 19 October 1916, 13.
84. *HMWC Interim Report*, p. 69.
85. J. M. Winter, *The Great War and the British People* (2nd edition, Basingstoke, 2003), pp. 204–245.
86. M. Baillie, 'The Red Women of Clydeside: Women Munition Workers in the West of Scotland During the First World War' (PhD Thesis, McMaster University, 2002).

87. The initial report was J. Campbell and L. E. Wilson, 'Inquiry into the Health of Women Engaged in Munition Factories', *HMWC Interim Report*, pp. 110–119. The later report was J. Campbell, 'Appendix B (I) A Further Inquiry into the Health of Women Munition Workers', *HMWC Final Report*, pp. 132–145.

88. 'Further Inquiry into the Health of Women Munition Workers', p. 139.

89. 'Work Fatigue – Shorter Hours to Increase Output', *The Times*, 21 December, 1918, 5.

90. *Annual Report of the Chief Inspector of Factories for the Year 1941*, p. 3.

91. *Annual Report of the Chief Inspector of Factories for the Year 1940*, p. 19.

92. Ibid, p. 20.

93. *Annual Report of the Chief Inspector of Factories for the Year 1941*, p. 23.

94. Ibid, p. 24.

95. *Annual Report of the Chief Inspector of Factories for the Year 1942*, pp. 5–8.

96. TUC Archive, MSS.292/147.6/1, 'War Time Rest Breaks for Industrial Workers', report enclosed in letter from W. H. Barton, Secretary of the Liverpool Trades Council and Labour Party to J. L. Smyth, TUC Social Insurance Secretary, 10 March 1942.

97. Ministry of Labour and National Service, 'Welfare Memorandum 3: Rest Breaks for Women and Girl Workers arranged by the Liverpool Union of Girls' Clubs', 14 November 1941.

98. Ministry of Labour and National Service Circular 128/87, 'Workers' Welfare: Holidays in Industry in 1942', 21 April 1942.

99. *Fighting Fit*, pp. 6–7.

100. TUC Archive, MSS.292/147.6/1, memorandum from William Elger, 'Industrial Welfare', 30 June 1941, copied to the TUC General Council.

101. *Annual Report of the Chief Inspector of Factories for the Year 1941*, p. 19.

102. *War Factory*, pp. 44–45.

103. *Annual Report of the Chief Inspector of Factories for the Year 1941*, p. 19.

104. For a fuller assessment of these schemes which interrogates their remit and impact, see N. Hayes, 'More Than "Music While You Eat"? Factory and Hostel Concerts, "Good Culture" and the Workers', in N. Hayes and J. Hill (eds), *Millions Like Us? British Culture in the Second World War* (Liverpool, 1999), pp. 209–235.

105. For example, *Annual Report of the Chief Inspector of Factories and Workshops for the Year 1911* (1912) Cd 6293, p. 159; see also *Annual Report of the Chief Inspector of Factories 1913*, p. 101.

106. The development of welfare supervision in this era is discussed in A. Wollacott, 'Maternalism, Professionalism and Industrial Welfare Supervisors in World War 1 Britain', *Women's History Review*, 3 (1994), 29–56.

107. *Annual Report of the Chief Inspector of Factories and Workshops 1918*, p. 43.

108. Woollacott, 'Maternalism, Professionalism and Industrial Welfare Supervisors', 38.

109. This organisation was renamed the Industrial Welfare Society in 1919, the periodical correspondingly changing to become the *Journal of Industrial Welfare* in 1920.

110. Sir William Beardmore, 'The Boys' Welfare Movement', *Boys' Industrial and Welfare Journal*, 1 (1918), 8.

111. *Annual Report of the Chief Inspector of Factories and Workshops for the Year 1918*, p. 46.
112. *Annual Report of the Chief Inspector of Factories for the Year 1941*, p. 9.
113. Wollacott, 'Maternalism, Professionalism and Industrial Welfare Supervision'.
114. *Annual Report of the Chief Inspector of Factories for the Year 1943*, pp. 56–59.
115. *Annual Report of the Chief Inspector of Factories for the Year 1941*, p. 10.
116. *Annual Report of the Chief Inspector of Factories for the Year 1943*, p. 56.
117. Ibid, pp. 56–59.
118. D. Lloyd George, Foreword, in Proud, *Welfare Work*, p. ix.
119. Ibid, p. xiii.
120. *Annual Report of the Chief Inspector of Factories and Workshops for the Year 1918*, p. iii.
121. Editorial 'Factory Work and Fatigue', *The Times*, 20 September 1915, 9: see footnote 17.
122. Editorial, 'The Human Machinery', *The Times*, 25 September 1920, 11.
123. *Annual Report of the Chief Inspector of Factories for the Year 1944*, p. 53.
124. *Annual Report of the Chief Inspector of Factories and Workshops for the Year 1919*, p. 112; *Annual Report of the Chief Inspector of Factories for the Year 1944*, p. 93; factories and workshops within Ireland were included for the final time in the figures for 1919 and their subsequent exclusion accounts for a small proportion of this decline.
125. Ibid, pp. 45–46.
126. Ibid, p. 44.
127. *Annual Report of the Chief Inspector of Factories for the Year 1940*, p. 23.
128. *Annual Report of the Chief Inspector of Factories for the Year 1943*, p. 41.
129. 'The Doctor in Dungarees', 88.
130. TUC Archive, MSS.292/141.2/5, minutes of the first meeting of the IHAC, 5 April 1943.

2 The Rise of the Healthy Factory

1. J. B. Priestley, *English Journey: Being a Rambling but Truthful Account of What One Man Saw and Heard and Felt and Thought During a Journey Through England During the Autumn of the Year 1933, Introduced by Margaret Drabble* (1934; London, 1997), p. 20.
2. See N. K. Buxton and D. H. Aldcroft (eds), *British Industry Between the Wars: Instability and Industrial Development 1919–1939* (London, 1979) for an overview and individual chapters devoted to the fate of specific industries.
3. T. W. Adorno and M. Horkheimer, *Dialectic of Enlightenment* (1944, trans. J. Cumming 1972, London, 1989), p. 131.
4. A. Marshall, *Principles of Economics* (1890), footnote 124; *Oxford English Dictionary*, (2nd edition, 1989), consulted online at http://dictionary.oed.com/ on 4 February 2008.
5. See D. Fitzgerald, *Every Farm a Factory: The Industrial Ideal in American Agriculture* (New Haven, 2003), pp. 106–128.
6. 'The British Association – Population and Food Supply: Sir Daniel Hall's Warning', *The Times*, 10 August 1926, 7.

7. D. Pick, *War Machine: The Rationalisation of Slaughter in the Modern Age* (Yale, 1993), p. 178.

8. Z. Bauman, *Modernity and the Holocaust* (Cambridge, 1999), p. 8, p. 22. Quoted in ibid, p. 187.

9. See V. Long, 'Industrial Homes, Domestic Factories: the Convergence of Public and Private Space in Interwar Britain', *Journal of British Studies*, 50 (2011). On the inspiration modernist architects took from industrial practices, see M. F. Guillén, *The Taylorized Beauty of the Mechanical: Scientific Management and the Rise of Modernist Architecture* (Princeton, 2006).

10. http://en.wikipedia.org/wiki/William_Morris, consulted on 1 February 2008.

11. W. Morris, 'Work in a Factory as it Might Be', *Justice*, 17 May 1884, 2; 'Work in a Factory as it Might Be II', *Justice*, 31 April 1884, 2; 'Work in a Factory as it Might Be III', *Justice*, 28 June 1884, 2.

12. Morris, 'A Factory as it Might Be', 17 May 1884.

13. Browne described his vision of the utopian asylum in the fifth lecture of his 1837 volume, *What Asylums Were, Are, and Ought to Be*, now most readily available in A. Scull (ed.), *The Asylum As Utopia: W. A. F. Browne and the Mid-Nineteenth Century Consolidation of Psychiatry* (London, 1991), pp. 176–231.

14. Ibid, p. 229.

15. E. Cumming and W. Kaplan, *The Arts and Crafts Movement* (London, 1991), p. 7, p. 18.

16. C. Harvey and J. Press, 'John Ruskin and the Ethical Foundations of Morris & Company, 1861–96', *Journal of Business Ethics*, 14 (1995), 181–194; 190.

17. D. Nye, *American Technological Sublime* (Cambridge, Massachusetts, 1994).

18. R. Williams, *The Country and the City* (London, 1973), p. 2.

19. W. W. Fowler, 'Gilbert White of Selbourne', *Macmillan's*, 68 (July 1893), pp. 182–189; cited in D. Worster, *Nature's Economy: A History of Ecological Ideas* (2nd edition, Cambridge, 1994), p. 15.

20. M. J. Weiner, *English Culture and the Decline of the Industrial Spirit 1850–1980* (Cambridge, 1981), p. 167.

21. D. Matless, *Landscape and Englishness* (London, 1998), pp. 25–102.

22. B. Luckin, *Questions of Power: Electricity and Environment in Interwar Britain* (Manchester, 1990), pp. 73–93.

23. See L. Briggs, *The Rational Factory: Architecture, Technology and Work in America's Age of Mass Production* (Baltimore and London, 1996).

24. E. Jones, *Industrial Architecture in Britain, 1750–1939* (London, 1985), p. 76.

25. Ibid; see also M. Stratton and B. Trinder, *Twentieth-Century Industrial Archaeology* (London, 2000), pp. 1–17.

26. Stratton and Trinder, *Twentieth-Century Industrial Archaeology*.

27. J. Skinner, *Form and Fancy: Factories and Factory Buildings by Wallis, Gilbert & Partners, 1916–1939* (Liverpool, 1997); a point also made in Stratton and Trinder, *Twentieth-Century Industrial Archaeology*.

28. Skinner, *Form and Fancy*, p. 264.

29. Ibid, pp. 270–271.

30. *Annual Report of the Chief Inspector of Factories and Workshops for the Year 1930* (1931) Cmd 3927, p. 66.

31. *Annual Report of the Chief Inspector of Factories and Workshops for the Year 1924* (1925), Cmd 2437, p. 7.
32. *Annual Report of the Chief Inspector of Factories and Workshops for the Year 1930*, p. 8.
33. *Annual Report of the Chief Inspector of Factories and Workshops for the Year 1937* (1938), Cmd 5802, p. 18.
34. For a Europe-wide introduction to health in the interwar years, see I. Borowy and W. D. Gruner (eds), *Facing Illness in Troubled Times: Health in Europe in the Interwar Years 1918–1939* (Frankfurt am Main, 2005). The most comprehensive account of developments in Britain is G. Jones, *Social Hygiene in Twentieth Century Britain* (London, 1986). In Jones' account, interwar health activities were viewed as part of a social hygiene movement, informed by social Darwinism, which was active between 1900 and 1960.
35. 'The Health Education Conference', *Public Health*, 44 (1930), 66–67.
36. C. Webster, 'Healthy or Hungry Thirties?', *History Workshop Journal*, 13 (1982), 110–129; H. L. Beales and R. S. Lambert (eds), *Memoirs of the Unemployed* (London, 1934).
37. See S. Sturdy, 'Hippocrates and State Medicine: George Newman Outlines the Founding Policy of the Ministry of Health', in C. Lawrence (ed.), *Greater than the Parts: Holism in Biomedicine 1920–1950* (Oxford, 1998), pp. 112–134.
38. Webster, 'Healthy or Hungry Thirties?'.
39. I. Zweiniger-Bargielowska, 'Raising a Nation of 'Good Animals': The New Health Society and Health Education Campaigns in Interwar Britain', *Social History of Medicine*, 20 (2007), 73–89.
40. David Matless provides an overview of mid-twentieth-century interest in diet before analysing proposals for an organic diet: see D. Matless, 'Bodies Made of Grass Made of Earth made of Bodies: Organicism, Diet and National Health in Mid-Twentieth-Century England', *Journal of Historical Geography*, 27 (2001), 355–376.
41. S. Carter, *Rise and Shine: Sunlight, Technology and Health* (Oxford, 2007).
42. J. Lewis and B. Brookes, 'The Peckham Health Centre, "PEP", and the Concept of General Practice During the 1930s and 1940s', *Medical History*, 27 (1983), 151–161; K. Barlow, 'The Peckham Experiment', *Medical History*, 29 (1985), 264–271.
43. For a detailed analysis of the Socialist Medical Association, see J. Stewart, *The Battle for Health: A Political History of the Socialist Medical Association* (Aldershot, 1999); H. Sigerist, *Socialised Medicine in the Soviet Union* (London, 1937).
44. *Annual Report of the Chief Inspector of Factories and Workshops for the Year 1911* (1912), Cd 6239, p. 46.
45. *Annual Report of the Chief Inspector of Factories and Workshops for the Year 1912* (1913) Cd 6852, p. 150.
46. *Annual Report of the Chief Inspector of Factories and Workshops for the Year 1913* (1914), Cd 7491, p. 65.
47. *Annual Report of the Chief Inspector of Factories and Workshops for the Year 1913*, p. 81.
48. Cited in D. Proud, *Welfare Work: Employers' Experiments For Improving Working Conditions in Factories with Foreword by David Lloyd George PM and first Minister of Munitions* (3rd edition, London, 1918), p. 38.

49. T. Oliver, *The Health of Workers* (London, 1925). Oliver's earlier principle publications had included *Diseases of Occupation from the Legislative, Social, and Medical Points of View* (1908), *Lead Poisoning in its Acute and Chronic Forms* (1891) and *Dangerous Trades* (1902).

50. Oliver, *The Health of Workers*, p. 28, 33.

51. E. W. Hope, in collaboration with W. Hanna and C. O. Stallybrass, *Industrial Hygiene and Medicine* (London, 1923), p. 45.

52. R. Squire, *Thirty Years in the Public Service: An Industrial Retrospect* (London, 1927), p. 45.

53. This is discussed in N. Tomes, *The Gospel of Germs: Men, Women and the Microbe in American Life* (Cambridge, Massachusetts, 1998), especially 'The Domestication of the Germ', pp. 135–156.

54. *Annual Report of the Chief Inspector of Factories and Workshops for the Year 1925* (1926), Cmd 2714, p. 39.

55. *Annual Report of the Chief Inspector of Factories and Workshops for the Year 1931* (1932), Cmd 4098, pp. 47–48.

56. *Annual Report of the Chief Inspector of Factories and Workshops for the Year 1936* (1937), Cmd 5514, p. 18.

57. First published in 1918 as the *Boys' Welfare Journal*, renamed the *Journal of Industrial Welfare* in 1920 and subsequently shortened to *Industrial Welfare* in 1922.

58. *Health of Munition Workers Committee Final Report: Industrial Health and Efficiency* (1918), Cd 9065, p. 106.

59. *Annual Report of the Chief Inspector of Factories and Workshops for the Year 1923* (1924), Cmd 2165, p. 80.

60. *Annual Report of the Chief Inspector of Factories and Workshops for the Year 1924*, pp. 85–86.

61. *Annual Report of the Chief Inspector of Factories and Workshops for the Year 1925*, pp. 96–98.

62. *Annual Report of the Chief Inspector of Factories and Workshops for the Year 1919* (1920), Cmd 941, p. 87.

63. *Annual Report of the Chief Inspector of Factories and Workshops for the Year 1937*, p. 78.

64. Modern Records Centre, University of Warwick, TUC Archive, MSS.292C/140/4, T. Legge, 'Prevention as a Benefit under the NHI Act', memorandum circulated to the TUC General Council, 24 June 1930.

65. See Chapter 3 for a political account of the 1937 Factories Act.

66. *Annual Report of the Chief Inspector of Factories and Workshops for the Year 1934* (1935), Cmd 4931, p. 51.

67. *Annual Report of the Chief Inspector of Factories and Workshops for the Year 1912*, p. 150.

68. The National Archives: Public Records Office, LAB 14/417, Home Office Industrial Museum Outline Guide, 1837–1961.

69. *Annual Report of the Chief Inspector of Factories and Workshops for the Year 1934*, p. 98.

70. For an analysis of the new approach adopted by the Inspectorate in the interwar years, see H. Jones, 'An Inspector Calls: Health and Safety at Work in Inter-war Britain', in P. Weindling (ed.), *The Social History of Occupational Health* (London, 1985), pp. 223–239.

71. *Annual Report of the Chief Inspector of Factories and Workshops for the Year 1932* (1933), Cmd 4377, p. 32.

72. TUC Archive, MSS.292/146.2/1, a letter from the TUC's Social Insurance Department to Malcolm Delevigne on 24 August 1928 indicated that the Industrial Museum had been featured in these publications.

73. *Annual Report of the Chief Inspector of Factories and Workshops for the Year 1932*, p. 83.

74. TUC Archive, MSS.292/146.2/1, letter from Mr L. Ward to J. L. Smyth, 28 January 1933.

75. *Annual Report of the Chief Inspector of Factories and Workshops for the Year 1934*.

76. *Annual Report of the Chief Inspector of Factories and Workshops for the Year 1937*, p. 88.

77. *Annual Report of the Chief Inspector of Factories and Workshops for the Year 1920* (1921), Cmd 1403, p. 70.

78. *Annual Report of the Chief Inspector of Factories and Workshops for the Year 1925*, p. 53.

79. *Annual Report of the Chief Inspector of Factories and Workshops for the Year 1930*, p. 68.

80. Jenson and Nicholson Ltd, *The Function of Colour in Factories Schools and Hospitals* (London, n.d., circa 1940s?), pp. 20, 40.

81. See Long, 'Industrial Homes'.

82. 'Welfare in a Congested Area, *Journal of Industrial Welfare* 2 (1920), 225–228; 227.

83. Photograph without parent article, captioned 'Ambulance room of Mather and Platt, Ltd. The large workshop can be seen through the window.' *Industrial Welfare* 6 (1924), 365. The photograph of the restroom at the Wallace Scott Tailoring Institute likewise depicts a room for which curtains cover the lower half of the windows, indicating that the adjoining space was in fact a production space: See C. U. Kerr, 'A Model Factory in Scotland', *Industrial Welfare* 6 (1924), 16–21; 19.

84. 'A New Scottish Welfare Building', *Industrial Welfare* 8 (1926), 335–339.

85. *Annual Report of the Chief Inspector of Factories and Workshops for the Year 1937*, p. 78.

86. Ibid, p. 77.

87. This is discussed in greater detail in Chapter 4.

88. *Annual Report of the Chief Inspector of Factories and Workshops for the Year 1931*, p. 75.

89. *Annual Report of the Chief Inspector of Factories and Workshops for the Year 1937*, p. 80. See also E. Robertson, M. Pickering and M. Korczynski, 'Harmonious Relations? Music at Work in the Rowntree and Cadbury Factories', *Business History*, 49 (2007), 211–234.

90. E. Hallas, 'The Trade Union Point of View', *Industrial Welfare*, 5 (1923), 31–34; 33.

91. *Annual Report of the Chief Inspector of Factories and Workshops for the Year 1930*, p. 64.

92. *Annual Report of the Chief Inspector of Factories and Workshops for the Year 1935* (1936), Cmd 5230, pp. 65–66.

93. For an analysis of the role played by working-class organisations in driving legislative change in this field, see S. Barton, *Working-Class Holidays and Popular Tourism, 1840–1970* (Manchester, 2005), pp. 107–132.

94. M. Glucksman, *Women Assemble: Women Workers and the New Industries in Inter-War Britain* (London, 1990), pp. 226–256.

95. P. Gurney, 'Labour's Great Arch: Cooperation and Cultural Revolution in Britain, 1795–1926', in E. Furlough and C. Strikwerda (eds), *Consumers Against Capitalism? Consumer Cooperation in Europe, North America, and Japan 1848–1990* (Oxford, 1999), pp. 135–171; p. 140. The interrelationship between the reform of consumption and of production is personified by Robert Owen who was renowned as both a pioneer of co-operative stores and a reformer of the factory system of production.

96. See M. Hilton, *Consumerism in Twentieth-Century Britain* (Cambridge, 2003), pp. 80–87. Though, as Gurney makes clear, wages and hours of workers employed in Co-operative Wholesale Society factories were rarely exemplary; mechanisation and the division of labour predominated and strikes were not uncommon: Gurney, 'Labour's Great Arch', p. 155.

97. Proud, *Welfare Work*, pp. 108–109.

98. B. Lewis, '*So Clean': Lord Leverhulme, Soap and Civilization* (Manchester, 2008), p. 116.

99. Beeston, Boots Company Archive, "Guide Handbook for the Beeston Factory," undated c. 1950s.

100. Proud, *Welfare Work*, pp. 182–183.

101. Ibid, pp. 108–109.

102. J. K. Walton, 'Towns and Consumerism', in M. Daunton (ed.), *The Cambridge Urban History of Britain Volume III 1840–1950* (Cambridge, 2000), pp. 715–744; p. 733. See also Lewis, *'So Clean'*, pp. 154–191.

103. See E. Robertson, *Chocolate, Women and Empire: A Social and Cultural History* (Manchester, 2009), pp. 9, 120–122.

104. See Long, 'Industrial Homes'.

105. For a discussion of the economic, social and political concerns which company towns were presented as the solution to, see M. Crawford, *Building the Workingman's Paradise: The Design of American Company Towns* (London, 1995). For specific British examples, see Lewis, *'So Clean'*, pp. 93–153. The close relationship between housing reform and industry in the interwar years is traced in Long, 'Industrial Homes'.

106. See F. H. A. Aalen, 'English Origins', in S. V. Ward (ed.), *The Garden City: Past, Present and Future* (London, 1992), pp. 28–51.

107. E. P. Cathcart, E. M. Bedale, C. Blair, K. Macleod and E. Weatherhead, *The Physique of Women in Industry – A Contribution Towards Determination of the Optimum Load: Industrial Fatigue Research Board Report no. 44* (London, 1927), p. 80.

108. S. V. Ward, 'The Garden City Introduced', in Ward, *The Garden City*, pp. 1–27.

109. Lewis, *'So Clean'*, p. 99.

110. Ibid, p. 94.

111. See C. Chin, *The Cadbury Story: A Short History* (Studley, 1998), pp. 19–35.

112. See H. Chance, 'The Angel in the Garden Suburb: Arcadian Allegory in the "Girls' Grounds" at the Cadbury Factory, Bournville, England, 1880–1930',

Studies in the History of Gardens and Designed Landscapes, 27 (2007), 197–216.

113. Birmingham Central Library, Cadbury Family Archives, MSS.466/8, 'The Factory in a Garden', four million ordered in June 1907; undated, untitled pamphlet, c. 1900s.

114. These trends mirrored advertising strategies adopted by Rowntrees which also drew upon local and national narratives of production while excluding details of its overseas cocoa producers. See Robertson, *Chocolate, Women and Empire*, p. 47.

115. Cadbury Family Archives, MSS.466/19, Cadbury, 'The Factory in a Garden', undated pamphlet c. 1910.

116. Cadbury Family Archives, MSS.466/8, black and white advertisement, circa 1890s; 'Bunny: Visit to Bournville', 'A Visit to Sunny Cocoa Land' and 'Bournville Bunny': all dated 7 June 1917 in a stock take of remaining supplies.

117. 'Life in Vickerstown – An Island Recreation Ground. How Barrow Workers are Housed', *The Times*, 6 August 1918, 9.

118. Emma Robertson has discussed how Rowntrees used images of clean white factories and white workers in white uniforms to represent hygiene and to symbolically cleanse products of their colonial origins: see Robertson, *Chocolate, Women and Empire*, p. 178. Similarly, Cadbury regularly featured images of its factory girls dressed in their white frocks within its advertising: Cadbury Family Archives, MSS.466/8.

119. Advertisement for Crosse and Blackwell, *The Times*, 8 June 1914, 4.

120. A. Adams, *Medicine by Design: the Architect and the Modern Hospital, 1893–1943* (Minneapolis, 2008), p. 109.

121. See R. Fitzgerald, *British Labour Management and Industrial Welfare, 1946–1939* (London, 1988).

122. See Chapter 4 for a discussion of the special attention paid to the needs of women working within industry.

123. Page advertisement for I. R. Morley, Textile Manufacturers, *The Times*, 25 June 1914, p. 6.

124. For example, advertisements placed by Chivers & Sons in *The Times* on 24 March 1925, 24 and 9 September 1925, 15; advertisements placed by J. Lyons & Co in *The Times* on 8 January 1925,1 6; 5 February 1925, 17 and 7 May 1925, 19; advertisement placed by Ovaltine in *The Times* on 16 February 1933, 8.

125. See M. Mayhew, 'The 1930s Nutrition Controversy', *Journal of Contemporary History*, 23 (1988), 445–464; C. Webster, 'Healthy or Hungry Thirties?', *History Workshop Journal*, 13 (1982), 110–129. Various aspects of the interwar debate on nutrition are also discussed by the contributors to David Smith's edited volume: D. F. Smith, *Nutrition in Britain: Science, Scientists and Politics in the Twentieth Century* (London, 1997).

126. British Medical Association, *Report of Committee on Nutrition* (London, 1933).

127. Advertisement, Ovaltine, *The Times*, 11 January 1926, 16.

128. T. Richards, *The Commodity Culture of Victorian England: Advertising and Spectacle, 1851–1914* (Stanford, 1990), p. 5.

129. Ibid., p. 30.

130. For a more detailed description of the design of the Boots factory, see Darley, *Factory*, pp. 122–128 and D. Yeomans and D. Cotton, *Owen Williams: The Engineer's Contribution to Contemporary Architecture* (London, 2001), pp. 82–92.

131. E. O. Williams, 'Factories – a Few Observations thereon made by Sir Owen Williams at a Discussion of the Art Workers Guild, 21 October, 1927', *Journal of the Royal Institute of British Architects*, 35 (1927), 54–55. Cited in ibid, p. 82.

132. For a rather uncritical account of these developments, see S. Phillips, 'Industrial Welfare and Recreation at Boots Pure Drug Company 1883–1945' (PhD thesis, Nottingham Trent University, 2003). Phillips lists the 1934 pamphlet as *Souvenir of Boots Beeston Factory*. The Owen Williams quotation is taken from p. 71 of the thesis.

133. Yeomans and Cottam, *Owen Williams*, p. 86.

134. Boots Company Plc, D122 Records Centre, Nottingham, *Achievement: a Record of Fifty Years of Progress of Boots Pure Drug Company* (1938), p. 47.

135. TUC Archive, MSS.292/140.9/1, report of visit to the Boots factory at Beeston by Dr Morgan, 1936.

136. *Annual Report of the Chief Inspector of Factories and Workshops for the Year 1914* (1915), Cd 8051, p. 1.

137. *Annual Report of the Chief Inspector of Factories and Workshops for the Year 1933* (1934). Cmd 4657, p. 78.

138. *Annual Report of the Chief Inspector of Factories and Workshops for the Year 1937*, p. 18.

139. N. Hayes, 'Did Manual Workers Want Industrial Welfare? Canteens, Latrines and Masculinity on British Building Sites, 1918–1970', *Journal of Social History*, 35 (2002), 637–658.

140. *Annual Report of the Chief Inspector of Factories and Workshops for the Year 1921* (1922), Cmd 1705, p. 64.

141. *Annual Report of the Chief Inspector of Factories and Workshops for the Year 1922* (1923), Cmd 1920, p. 57.

142. J. Melling, 'Employers, Industrial Welfare, and the Struggle for Work-Place Control in British Industry, 1880–1920', in H. Gospel and C. R. Littler (eds), *Management and Labour in British Business Strategies*, (London, 1983), pp. 55–81; H. Jones, 'Employers' Welfare Schemes and Industrial Relations in Inter-War Britain', *Business History*, 25 (1983), 61–75.

143. TUC Archive, MSS.292C/146.9/6, TUC General Council, 'The Effect of Welfare Work on Trade Unionism', typescript, 24 May 1932.

144. A. Meharg, 'The Arsenic Green', *Nature*, 423 (2003), 688.

145. Certifying factory surgeons undertook statutory investigations into notifiable cases of industrial disease and industrial accidents, statutory examinations of workers engaged in scheduled dangerous processes and of young workers entering industrial employment to certify their fitness to work. This group of doctors were re-designated appointed factory doctors under the 1937 Factory Act, most worked primarily as general practitioners.

146. This receives further consideration in Chapter 3 where endeavours to improve the position of industrial medicine will be studied.

147. Nikolas Rose has argued that psychological factors displaced physiological explanations for fatigue and inefficiency in industry in the 1920s and

30s: N. Rose, 'The Contented Worker', in *Governing the Soul: The Shaping of the Private Self* (2nd edition, London, 1999), pp. 61–75.

148. For a discussion on the development of industrial psychology in Britain and its objectives see M. Thomson, 'Psychology and the Problem of Industrial Civilisation', in *Psychological Subjects: Identity, Culture and Health in Twentieth-Century Britain* (Oxford, 2006), pp. 140–172.

3 Taking Responsibility: The Politics of Industrial Health

1. Modern Records Centre, University of Warwick (hereafter MRC), TUC Archive, MSS.292/140/2, letter from H. E. Collier to H. Morgan, 4 July 1944.
2. Ibid, letter from H. Morgan to H. E. Collier, 19 July 1944.
3. J. Habermas, *The Structural Transformation of the Public Sphere* (1962, trans. T. Burger 1989, Cambridge, 1999).
4. K. Middlemas, *Politics in Industrial Society: The Experience of the British System Since 1911* (London, 1979).
5. N. Fraser, 'Rethinking the Public Sphere: A Contribution to the Critique of Actually Existing Democracy', in C. Calhoun (ed.), *Habermas and the Public Sphere* (1992, Cambridge, Massachusetts, 1999), pp. 109–142; p. 132. Italics used in original text.
6. S. Sturdy (ed.), *Medicine, Health and the Public Sphere in Britain, 1600–2000* (London, 2002).
7. Ibid, p. 20.
8. P. Bourdieu, *The Logic of Practice* (Cambridge, 1990, translated by R. Nice).
9. H. Perkins, *The Rise of Professional Society: England Since 1880* (London, 1990).
10. Ibid, pp. 1–17.
11. J. Melling, 'The Risks of Working and the Risks of Not Working: Trade Unions, Employers and Responses to the Risk of Occupational Illness in British Industry, c. 1890–1940s', *ESRC Discussion Paper No. 12* (2003), pp. 14–34.
12. Ibid, p. 16.
13. Arthur McIvor's study of employer associations in Northern England points to a shift from collective action to regulation on a company level in the interwar years: see A. McIvor, *Organised Capital: Employers' Associations and Industrial Relations in Northern England, 1880–1939* (Cambridge, 1996).
14. On the psychological impact of unemployment, see H. L. Beales and R. S. Lambert (eds), *Memoirs of the Unemployed* (London, 1934) and R. McKibbin, *The Ideologies of Class: Social Relations in Britain, 1880–1950* (Oxford, 1991), pp. 228–258. On nutrition, see British Medical Association, *Report of Committee on Nutrition* (London, 1933). See M. Mayhew, 'The 1930s Nutrition Controversy', *Journal of Contemporary History*, 23 (1988), 445–464, also C. Webster, 'Healthy or Hungry Thirties?', *History Workshop Journal*, 13 (1982) 110–129.
15. For an analysis of the State's relationship with labour organisations and the legislation passed during the course of the First World War to regulate work and minimise strike action, see G. R. Rubin, *War, Law and Labour: The*

Munitions Acts, State Regulation, and the Unions, 1915–21 (Oxford, 1987). For a brief analysis of the TUC's role in these events, see D. F. MacDonald, *The State and the Trade Unions* (2nd edition, London, 1976), pp. 82–96.

16. MacDonald, *The State and the Trade Unions*, p. 82.
17. See J. Hinton, *The First Shop Stewards' Movement* (London, 1973).
18. See 'Walter Citrine and TUC Modernisation, 1926–1939', in R. Taylor, *The TUC: From the General Strike to New Unionism* (Houndmills, 2000), pp. 20–75 for a more detailed overview of the expansion and professionalisation of the TUC in this era. Also 'Trade Unions and the Depression', in MacDonald, *The State and the Trade Unions*, pp. 97–117.
19. Taylor, *The TUC*, pp. 27–75.
20. Known as the Mond–Turner talks, after the 1928 TUC President Ben Turner and Sir Alfred Mond, the chairman of Imperial Chemical Industries. See Taylor, *The TUC*, pp. 44–48.
21. The TUC's archives, now held at the Modern Records Centre, University of Warwick, are still organised by their original classificatory system. Catalogued under broad subject areas, including labour conditions, industrial relations, politics and government and social issues, these files enable historians to trace the TUC's growing interest in a broad range of social and political issues.
22. W. Milne-Bailey, *Trade Unions and the State*, (London, 1934).
23. Taylor, *The TUC*, p. 74.
24. MacDonald, *The State and the Trade Unions*, p. 97 and p. 113.
25. Trade union membership stood at 4.8 million in 1930. Membership and density statistics from G. Bain and R. Price, *Profiles of Union Growth* (Oxford, 1980), pp. 37–38, cited in A. McIvor, *A History of Work in Britain, 1880–1950* (Basingstoke, 2001), p. 201; see pp. 200–240 for a detailed analysis of trade union strength in the workplace and at the level of national politics.
26. Taylor, *The TUC*.
27. Discussed in ibid, pp. 76–101; p. 76.
28. H. Jones, 'Employers' Welfare Schemes and Industrial Relations in Inter-War Britain', *Business History*, 25 (1983), 61–75; P. W. J. Bartrip, *Workmen's Compensation in Twentieth Century Britain: Law, History and Social Policy* (Aldershot, 1987).
29. *Interim Report on the Future Provision of Medical and Allied Services* (1920), Cmd 693, discussed in C. Webster, *The Health Services Since the War Volume 1: Problems of Health Care – The National Health Service Before 1957* (London, 1988), p. 19.
30. Dawson, 'Medicine and the State', *The Medical Officer*, 23 (1920), 223–224; cited in Webster, *The Health Services*, p. 19.
31. For an analysis of the BMA's strategy in this era, see P. Bartrip, *Themselves Writ Large: The BMA 1832–1966* (London, 1996) and R. Cooter, 'The Rise and Decline of the Medical Member: Doctors and Parliament in Edwardian and Interwar Britain', *Bulletin for the History of Medicine*, 78 (2004), 59–107.
32. Lord Hill of Luton, *The Other Side of the Hill* (London, 1964), p. 97; cited in Bartrip, *Themselves Writ Large*, p. 161.
33. Cooter, 'The Rise and Decline of the Medical Member', p. 79.
34. Ibid, p. 61.
35. BMA, *Report of the Nutrition Committee*.

36. TUC Archive, MSS.292C/140.9/3, H. B. Morgan, 'Medical Evidence and the Proposed New Medical Curriculum', memorandum produced for the TUC General Council, 2 February 1936; TUC General Council, 'Trade Union representation on Quasi-Medical Bodies', memorandum for the consideration of the Workmen's Compensation and Factories Committee, 22 February 1935.

37. For an account of the Socialist Medical Association, see J. Stewart, *'The Battle for Health': A Political History of the Socialist Medical Association, 1930–51* (Aldershot, 1999); for an analysis of the medical politics of the 1930s from the perspective of the SMA, including the interactions of the TUC, BMA, Labour Party and Medical Practitioners Union, see pp. 62–84.

38. Founded as the Medico-Political Union in 1914, the union became the Medical Practitioners' Union in 1922.

39. For a more detailed account of the relationship between the Labour Party, the trade unions and the TUC in the interwar years, see J. Hinton, *Labour and Socialism: A History of the British Labour Movement 1867–1974* (Brighton, 1983), pp. 119–160.

40. TUC Archive, MSS.292C/140/4, T. Legge, draft memoranda produced for the TUC General Council, 'Prevention as a Benefit under the NHI Act', 24 June 1930.

41. See G. Finlayson, *Citizen, State, and Social Welfare in Britain 1830–1990* (Oxford, 1994), pp. 201–286 for an analysis of role of voluntary organisations in the interwar years.

42. TUC Archive, MSS.292/140.9/1, memorandum, 'Medical Advisor', circa 1932.

43. Somerville Hastings, *Lancet*, I (1933), 1324 and 'The First Steps Towards a Socialized Medical Service', *Medicine Today and Tomorrow*, July 1938, 4: Cited in Stewart, *The Battle for Health*, p. 41.

44. TUC Archive, MSS.292/140.9/1, memorandum, 'Duties of Medical Advisor', 23 October 1940.

45. For a brief overview of Legge's career, see P. W. J. Bartrip, 'Legge, Sir Thomas Morison (1863–1932)', *Oxford Dictionary of National Biography* (Oxford, 2004), www.oxforddnb.com/view/article/49286, accessed 15 April 2008.

46. TUC Archive, MSS.292/140.9/1, 'Industrial Disease with special reference to Silicosis': speech given by T. Legge to the National Council of the Pottery Industry, 24 March 1930.

47. T. Legge, *Industrial Maladies* (Oxford, 1934).

48. TUC Archive, MSS.292/140.9/1, memorandum, 'Medical Advisor', circa 1932.

49. 'Obituary: Dr H. B. W. Morgan – TUC Medical Advisor', *The Times*, 9 May 1956, 13.

50. TUC Archive, MSS.292/844.8/1, letter from Dr H. Morgan to Dr C. Hill, 19 October 1935.

51. Ibid, letter from Dr Hill to Dr Morgan, 16 October 1935.

52. Ibid, letter from Morgan to Hill, 19 October 1935.

53. 'Obituary: Dr H. B. W. Morgan', 13.

54. TUC Archive, MSS.292/140/1, H. Morgan, 'Sickness in Industry', unpublished letter to *The Times*, sent 7 November 1934.

55. TUC Archive, MSS.292C/140.9/3, Workmen's Compensation and Factories Committee, 'Medical Evidence and the Proposed New Medical Curriculum', memorandum, 20 February 1936.

56. TUC Archive, MSS.292/140/2, H. Morgan, 'Industrial Medicine', memorandum produced for W. Citrine, 5 October 1944.
57. TUC Archive, MSS.292/140/1, flyer from University of Birmingham, 'Department of Industrial Hygiene and Medicine – Reader Howard E. Collier', undated, circa 1934/1935.
58. TUC Archive, MSS.292/140/2, letter from H. E. Collier to H. Morgan, 19 July 1944.
59. TUC Archive, MSS.292/140/2, H. Morgan, 'Industrial Medicine', memorandum produced for W. Citrine, 5 October 1944.
60. This incident is documented in Stewart, *The Battle for Health*, pp. 185–186.
61. TUC Archive, MSS.292/844.8/1, letter from the South Wales Miners Federation to the TUC, 12 April 1935.
62. TUC Archive, MSS.292/844.8/1, 'Dr Summerskill – Prediction Concerning Local Workmen's Medical Scheme', 11 April 1935; 'Llanelly and District Workmen's Medical Council: Honour Mr Morgan J. P. after 25 Years' Service', 11 April 1935 and 'New Battle Between Doctors', 25 April 1935: all from the *Llanelly and County Guardian*. Articles enclosed in letter sent from Fred Smith to the TUC, 22 May 1935.
63. TUC Archive, MSS.292/844.8/1, letter from Smyth to Evans, 28 June 1935.
64. TUC Archive, MSS.292/844.8/1, 'Llanelly Medical Dispute', typescript prepared by Dr Morgan, 2 November 1935.
65. TUC Archive, MSS.292/844.8/1, memorandum of interview with Dr Anderson on the subject of the BMA, 24 April 1935.
66. TUC Archive, MSS.292/844.8/1, letter from H. Morgan to J. L. Smyth, 16 November 1935.
67. See subsequent files, TUC Archive, MSS.292/844.1/2-3.
68. TUC Archive, MSS.292/844.8/3, letter from J. L. Evans to Dr Morgan, March 2 1948.
69. R. Earwicker, 'A Study of the BMA–TUC Joint Committee on Medical Questions, 1935–1939', *Journal of Social Policy*, 8 (1979), 335–561; 336.
70. TUC Archive, MSS.292/840/3, BMA / TUC Joint Committee Meeting Minutes. For a more detailed analysis of this Committee, and in particular an assessment of the maternity scheme developed by the Committee, see Earwicker, 'A Study of the BMA–TUC Joint Committee'.
71. TUC Archive, MSS.292/844.1/4, letter from Dr Brock to Smyth, 2 December 1930; memorandum of an interview between Dr Bushnell and Smyth, 13 January 1932.
72. TUC Archive, MSS.292/844.1/4, 'SMA Report of the Research Sub-Committee', typescript, 1933.
73. TUC Archive, MSS.292/140/2, letter from A. W. Bourne to the TUC, 21 January 1942; handwritten comments by Morgan appended to the note.
74. TUC Archive, MSS.292/841.1/4, Tewson wrote to the unions instructing them not to participate on 17 July 1947; the Secretary of the SMA wrote to the TUC on 30 July 1947 and the cited response to the SMA was sent by Tewson on 20 August 1947.
75. Ibid, letter from the Secretary of the Organisation Department of the TUC to Alderman A. A. Rignall, secretary of the Battersea Labour Party and Trades Council, 3 October 1950.

76. Ibid, letter sent by the Trades Councils Section of the TUC to Mr R. Sanders of the Swansea and District Trades Council, August 1950.

77. Ibid, letter from V. Tewson to M. Philips, 9 November 1951; letter from Dr S. Murray to M. Philips, 21 November 1951.

78. Ibid, letter from Tewson to M. Philips, 8 January 1952.

79. Ibid, interdepartmental correspondence from C. R. Dale to V. Tewson, 29 February 1952. This comment was crossed out but remained legible.

80. Discussed in Stewart, *The Battle for Health*, pp. 72–73.

81. J. Lee, *The Principles of Industrial Welfare* (London, 1924), p. 2. John Lee was the Controller of the Central Telegraph Office and the first Editor of the *Journal of Public Administration*.

82. TUC Archive, MSS.292C/146.9/6, TUC General Council, 'The Effect of Welfare Work on Trade Unionism', 24 May 1932, typescript report.

83. J. Melling, 'Employers, Industrial Welfare, and the Struggle for Work-Place Control in British Industry, 1880–1920', in H. Gospel and R. C. Littler (eds), *Management and Labour in British Business Strategies* (London, 1983), pp. 55–81.

84. For more detail on the nature of mental health problems believed to be experienced by workers, see Chapter 4. These memoranda and files have also been discussed by Mathew Thomson, who argued that the TUC's cautious response was in part engendered by an unwillingness to admit that workmen's health problems could have a psychological rather than organic basis: See M. Thomson, *Psychological Subjects: Identity, Culture, and Health in Twentieth-Century Britain* (Oxford, 2006), pp. 167–171.

85. TUC Archive, MSS.292/140.1/2, Dr M. Rosenfield, 'Industrial Mental Hygiene Clinics', *Industrial Welfare and Personnel Management* (February 1935), reprint.

86. TUC Archive, MSS.292/140.1/2, letter from Rosenfield to Smyth, 28 May 1935.

87. TUC Archive, MSS.292/140.1/2, memorandum produced by Rosenfield for the TUC, 27 July 1935.

88. TUC Archive, MSS.292/140.1/2, internal memorandum from J. L. Smyth to H. Morgan, 12 March 1935.

89. TUC Archive, MSS.292/140.1/2, J. R. Rees, 'The Problem of Neurotic Disorder in its Relation to Industrial Efficiency', undated.

90. TUC Archive, MSS.292/140.1/2, J. W. Yerrell, 'Memorandum on Nervous Diseases', October 1937.

91. TUC Archive, MSS.292/140.1/2, Dr E. N. Snowden, 'Injury Psychoneurisis (Traumatic Neurasthenia)', 3 January 1938.

92. For an examination of the significance of fractures and orthopaedics in the politics of interwar medicine, including an analysis of the interest displayed by the TUC, see R. Cooter, 'The Meaning of Factures: Orthopaedics and the Reform of British Hospitals in the Inter-War Period', *Medical History*, 31 (1987), 306–332.

93. TUC Archive, MSS.292/842.1/1-3, Manor House Labour Hospital 1925–36; 1937–46; 1947–53. The main records for the hospital are held at the London Metropolitan Archives.

94. Although many friendly societies acted as approved societies following the introduction of the National Health Insurance Act in 1911, Joan Lane

argued that their assumption of statutory responsibilities transformed the character of the societies, undermining their public role and ethos of mutuality: the longevity of the Industrial Orthopaedic Society, which urged unions and individuals to donate to provide a service above and beyond statutory provisions clearly went against this general trend. See J. Lane, *A Social History of Medicine: Health, Healing and Disease in England, 1750–1950* (London, 2001), p. 79.

95. TUC Archive, MSS.292/842.1, 'The First Labour Hospital: Workers of the World Unite', undated leaflet circa 1920s.
96. TUC Archive, MSS.292/842.2: these beds were endowed in 1942, 1944 and 1945 respectively.
97. TUC Archive, MSS.292/842.1, TUC General Council memoranda, 'Manor House Hospital and a Ward for Industrial Diseases', undated. This appears to date from the late 1930s before the TUC decided to endow the single-bed ward instead.
98. TUC Archive, MSS.292/140/1, interdepartmental memorandum from H. Morgan to J. L. Smyth, 'Industrial Health Lectures and Trade Unions', 6 July 1939.
99. TUC Archive, MSS.292/842.2/1, article from the *Preston Gazette*, 21 March 1931, 9.
100. TUC Archive, MSS.292/842.2/1, letter from Citrine to the secretaries of all affiliated unions, 16 April 1932.
101. This organisation was initially named the Industrial Educational Council, re-titled the Industrial Health Education Council in July 1925 and becoming the Industrial Health Education Society in July 1926, but to avoid unnecessary confusion I have simply referred to it as the IHES throughout.
102. TUC Archive, MSS.292/141.1/2, pamphlet for the Industrial Education Council dated March 1925.
103. TUC Archive, MSS.292/141.1/2, TUC General Council, report of Mr Mackenzie's statement respecting the work of the IHES, 11 February 1927.
104. TUC Archive, MSS.292/141.1/1, pamphlet produced by the Industrial Education Council, 1925.
105. TUC Archive, MSS.292/141.1/2, 'The Health of the Worker', typescript article ascribed to Ben Tillett. Letter from A. S. Firth to Tillett, 22 October 1928.
106. TUC Archive, MSS.292/141.1/2, J. Johnstone Jervis, 'The Healthy Worker', typescript.
107. TUC Archive, MSS.292/141.1/2, memorandum of interview between Citrine, Smyth and Mackenzie, 7 October 1929; letter from Mackenzie to Citrine, 15 October 1929.
108. 'The Public Health Department. From the Desk of a "M.o.H." The Industrial Health Education Society', *Municipal Engineering, Sanitary Record and Municipal Motor*, 7 November 1929, 520.
109. TUC Archive, MSS.292/141.1/3, inter-departmental TUC correspondence from Legge to Smyth, 24 February 1930.
110. TUC Archive, MSS.292/141.1/3, inter-departmental TUC correspondence from Legge to Smyth, 24 May 1930.

111. TUC Archive, MSS.292/141.1/3, inter-departmental correspondence from Herbert Tracey to Legge, 5 March 1930; Legge to Tracey, 5 March 1930; Tracey to Citrine, 6 March 1930.
112. This is discussed at length in Chapter 5.
113. TUC Archive, MSS.292/141.1/3, IHES memorandum for speakers, 11 March 1930.
114. TUC Archive, MSS.292/141.1/3, memorandum from Legge regarding the IHES Executive Council meeting held on 22 July 1930, sent 24 July 1930.
115. TUC Archive, MSS.292/141.1/3, undated letter written by Legge to the Executive Committee of the IHES; Letter from Sydney Walton to Legge, 20 March 1930.
116. TUC Archive, MSS.292/141.1/3, TUC General Council, 'IHES – Notes on Proceedings at the Committee of Enquiry held at BMA House, Friday 28th November 1920', 29 November 1930.
117. TUC Archive, MSS.292/1/141.1/3, letter from Citrine to Mackenzie, 7 July 1932, emphasis added.
118. TUC Archive, MSS.292/141.1/4, letter from Mackenzie to H. B. Morgan, 12 April 1933.
119. TUC Archive, MSS.292/141.1/4, memoranda produced by Morgan on the IHES, 30 August 1933; 7 October 1933; 8 November 1933, 26 February 1934.
120. TUC Archive, MSS.292/141.1/4, report of a meeting between Mackenzie and the TUC Workmen's Compensation and Factories Committee, 11 July 1934.
121. TUC Archive, MSS.292/141.1/5, letter from Mr W. M. Emerson to Mr E. P. Harries, Secretary of the TUC General Council, 1935.
122. TUC Archive, MSS.292/141.1/5, letter from Smyth to Citrine, 2 July 1935.
123. See discussion on employers, workers and welfare in previous chapter.
124. MRC, BEC Archive, MSS.200/B/3/2/C693 pt 3, letter from Adam Nimmo to Forbes Watson, 10 January 1928.
125. TUC Archive, MSS.292/141.1/6, letter from Mackenzie to Smyth, 28 September 1937.
126. TUC Archive, MSS.292C/140/1, W. Citrine, 'Health Education', memorandum, 6 February 1935.
127. TUC Archive, MSS.292C/140/1, H. B. Morgan, 'Health Education', memorandum for the TUC General Council, 19 February 1935 and H. B. Morgan, 'Health Advisory Committee', memorandum for the TUC General Council, 31 May 1935.
128. TUC Archive, MSS.292C/140/1, H. Morgan, 'General Health Committee', memorandum, 5 June 1935.
129. A further letter sent by Morgan to Smyth enquiring as to whether there had been any developments regarding the establishment of the proposed committee elicited no response: TUC Archive, MSS.292C/140/1, 'Proposed New Health Committee', inter-departmental correspondence from Morgan to Smyth, 29 November 1935.
130. TUC Archive, MSS.292C/140.9/3, Holidays with Pay Committee, 'Committee of Enquiry and TUC Evidence', memorandum, 25 May 1937. It is noteworthy that this campaign was pursued by seeking legislative change, reflecting a change in strategy.

131. TUC Archive, MSS.292/ 141.1/6, memorandum regarding the IHES Financial Committee meeting, 24 January 1938.
132. TUC Archive, MSS.292/141.2/3, minutes of an ad hoc committee of the Central Council for Health Education on industrial health, 5 March 1948.
133. TUC Archive, MSS.292/141.2/3, meeting of an ad hoc committee of the Central Council for Health Education on general health education at the place of work, 20 April 1948.
134. TUC Archive, MSS.292/141.2/3, meeting of the Social Insurance Committee, 13 May 1948.
135. TUC Archive, MSS.292/141.2/3, letter from Sutherland to the Social Insurance Committee, 10 June 1949. Conversely, in the post-war context the TUC would become critical of the perception that the employee in the workplace was a worker but not a citizen, and would seek to reassert the specific industrial health problems which undermined the health of the employee: this is explored in Chapter 6.
136. H. Jones, 'An Inspector Calls: Health and Safety at Work in Inter-War Britain', in P. Weindling (ed.), *The Social History of Occupational Health* (London, 1985), pp. 223–239.
137. *Annual Report of the Chief Inspector of Factories and Workshops for the Year 1937* (1938), Cmd 5802, p. 11.
138. *Annual Report of the Chief Inspector of Factories and Workshops for the Year 1930* (1931), Cmd 3297, p. 11.
139. More details of what the Act entailed are given in the previous chapter.
140. National Archives, Public Records Office, LAB 14/57, memorandum on the Factories Bill prepared in preparation for TUC deputation, 30 January 1936.
141. TUC Archive, MSS.292C/140/1, TUC Research Department, Legge 'Factories Bill', typescript dated 1 February 1927.
142. Public Records Office, LAB 14/57, report of deputation to the Secretary of State for Home Affairs from the General Council of the TUC, 30 January 1936.
143. TUC Archive, MSS.292C/140/4, medical inspection 1912 and 1925–1931, reports of Sir Thomas Legge.
144. *Annual Report of the Chief Inspector of Factories and Workshops for the Year 1937*, p. 7.
145. *Annual Report of the Chief Inspector of Factories and Workshops for the Year 1920* (1921), Cmd 1402, p. 115.
146. TUC Archive, MSS.292C/140/4, T. Legge, 'Prevention as a Benefit under the NHI Act', memorandum produced for the TUC, 24 June 1930.
147. TUC Archive, MSS.292C/140.9/3, 'Proposed Extension of National Health Insurance Benefits to Uninsured Persons', memorandum submitted to the TUC General Council, 12 March 1936.
148. TUC Archive, MSS.292C/140/4, Legge, 'Prevention as a Benefit under the NHI Act', memorandum produced for the TUC, 24 June 1930.
149. This is discussed at length in Chapter 5.
150. TUC Archive, MSS.292/134.1/5, excerpt from deputation to Minister of Health, contained within a summary produced by the TUC Women's Advisory Committee on the health of women in industry, 9 March 1939.

151. See discussion in the previous chapter on the respective merits of home and work environments on workers' health: this topic was also crucial in the post-war debates regarding the shape of an industrial health service, discussed in Chapter 5.

152. TUC Archive, MSS.292/140/2, letter from Morgan to Collier, 19 July 1944.

153. TUC Archive, MSS.292/140/2, letter from H. Morgan to Dr B. Stross, 25 March 1941.

154. P. Weindling, 'Linking Self Help and Medical Science: The Social History of Occupational Health', in Weindling (ed.), *The Social History of Occupational Health*, pp. 2–31; p. 10.

155. Bartrip, *Workmen's Compensation*, pp. 88–167.

156. Arthur McIvor and Ronald Johnston have contrasted the relative ease with which coal workers' pneumoconiosis was scheduled as an industrial disease when compared with emphysema and bronchitis, diseases not confined to coal workers: See A. McIvor and R. Johnston, *Miner's Lung: A History of Dust Disease in British Coal Mining* (Aldershot, 2007), pp. 123–142.

157. P. Bartrip, *The Home Office and the Dangerous Trades: Regulating Occupational Disease in Victorian and Edwardian Britain* (Amsterdam and New York, 2002), pp. 276–277.

158. S. Bowden and G. Tweedale, 'Mondays without Dread: The Trade Unions Response to Byssinosis in the Lancashire Cotton Industry in the Twentieth Century', *Social History of Medicine*, 16 (2003), 79–96; 94–95.

159. TUC Archive, MSS.292/140.1/2, letter from Dr M. Rosenfield to Smyth, 28 May 1935.

4   Tailoring Provisions for Individualised Needs

1. E. T. Kelly (ed.), *Welfare Work in Industry – By Members of the Institute of Welfare Workers* (London, 1925), p. 1: capitalisation in original text.

2. This organisation was first established in 1917 as the Central Association for Welfare Workers; it was subsequently renamed the Institute of Industrial Welfare Workers Incorporated (1924), the Institute of Labour Management (1931), the Institute of Personnel Management (1946), the Institute of Personnel and Development (1994) and the Chartered Institute of Personnel and Development (2000).

3. These tendencies should not be overstressed. Arthur McIvor argues that while labour intensification was fairly pervasive between 1880 and 1950 as British companies struggled to compete in an increasingly competitive international market, job fragmentation and deskilling was a less uniform process, curtailed by worker resistance, the unwillingness of employers to inaugurate changes when a ready supply of labour existed and the creation of new skilled positions in those fields of work that did undergo deskilling. See A. McIvor, *A History of Work in Britain, 1880–1950* (Basingstoke, 2001), pp. 43–78.

4. See D. F. MacDonald, *The State and the Trade Unions* (2nd edition, London, 1976), pp. 43–52; also J. Hinton, *Labour and Socialism: A History of the British Labour Movement 1867–1974* (Thetford, 1983), pp. 40–82.

5. The term 'industrial warfare' was first used in *The Times* in 1876 in a report of the annual meeting of the National Federation of Employers of Labour, at which it was asserted that as the trade unions had formed a federation it was time for the employers to do likewise, 'not to attack Trades Unions, but for their own protection': 'Employers and Trade Unions', *The Times*, 26 February 1876, 5. The phrase 'industrial warfare' was used in *The Times* three times in the 1870s, five times in the 1880s, 31 times in the 1890s, 20 times in the 1900s, 44 times in the 1910s and was last mentioned in 1920, when it was used four times.

6. H. Duberry, 'What a Trade Unionist Thinks of Welfare Work', *Journal of Industrial Welfare*, 2 (1920), 12–13; 12. Duberry was the General Secretary of the Post Office Supervising Trade Union.

7. J. Drever, 'The Human Factor in Industrial Relations', in C. S. Myers (ed.), *Industrial Psychology* (2nd edition, Oxford, 1944: first published 1929), pp. 16–38; p. 17.

8. For an overview of the development of company welfare schemes in different industrial sectors, see R. Fitzgerald, *British Labour Management and Industrial Welfare, 1846–1939* (London, 1988). Fitzgerald argued that systematic welfare provisions assumed a new importance in increasingly large and complex organisations, serving to minimise labour turnover and industrial discontent. Among smaller firms with lower profit margins, discretionary welfare provisions continued to predominate.

9. S. Sturdy, 'The Industrial Body', in R. Cooter and J. Pickstone (eds), *Companion to Medicine in the Twentieth Century* (London, 2003), pp. 217–234.

10. This transformation is discussed in Chapters 1 and 2.

11. N. Rose, *Governing the Soul: The Shaping of the Private Self* (2nd edition, London, 1999).

12. 'The Subject at Work', 'The Contented Worker', 'The Worker at Work', 'Democracy at Work' and 'The Expertise of Management', in ibid, pp. 55–102.

13. See, for example, J. Melling, 'Employers, Industrial Welfare, and the Struggle for Work-Place Control in British Industry, 1880–1920', in H. Gospel and C. R. Littler (eds), *Management and Labour in British Business Strategies* (London, 1983), pp. 55–81.

14. Rose, *Governing the Soul*, p. 108.

15. Rose argued that the wage relation was 'encumbered' by health and safety legislation: as discussed in Chapter 2, the provisions contained within the 1901 Factory Act to govern workplace conditions were in practice minimal. Furthermore, Rose asserted that the destiny of industrial welfare workers 'was to be profoundly affected by World War 1', a statement which misleadingly implies that there was a definable body of welfare supervisors before the First World War: ibid, p. 64.

16. Ibid, p. 98.

17. See also M. Glucksman, 'Homeward Bound: Changes in Domestic Production and Consumption' in *Women Assemble: Women Workers and the New Industries in Inter-War Britain* (London, 1990), pp. 226–256; V. Long, 'Industrial Homes, Domestic Factories: the Convergence of Public and Private Space in Interwar Britain', *Journal of British Studies*, 50 (2011).

18. Rose, *Governing the Soul*, p. 105.

19. W. Morris, 'Work in a Factory as it Might Be', *Justice*, 17 May 1884, 2; 'Work in a Factory as it Might Be II', *Justice*, 31 April 1884, 2; 'Work in a Factory as it Might Be', *Justice*, 28 June 1884, 2. Discussed in Chapter 2.
20. M. Thomson, *Psychological Subjects: Identity, Culture and Health in Twentieth-Century Britain* (Oxford, 2006).
21. Ibid, p. 140.
22. Ibid, p. 142.
23. M. Jahoda, P. F. Lazarsfeld and H. Zeisel, *Marienthal: The Sociology of an Unemployed Community* (London, 1974. First published as *Die Arbeitslosen von Marienthal* in 1933), p. 19.
24. H. L. Beales and R. S. Lambert (eds), *Memoirs of the Unemployed* (London, 1934).
25. 'The "Social Psychology" of Unemployment in Inter-War Britain', in R. McKibbin, *The Ideologies of Class: Social Relations in Britain, 1880–1950* (Oxford, 1991), pp. 228–258.
26. M. Johnstone, 'John Gollan (1911–1977)', *Oxford Dictionary of National Biography* (Oxford, 2004), www.oxforddnb.com/view/article/31156, accessed 20 May 2008.
27. J. Gollan, *Youth in British Industry: A Survey of Labour Conditions To-day* (London, 1937), p. 155.
28. G. Meara, *Juvenile Unemployment in South Wales* (Cardiff, 1936), cited in Gollan, *Youth in British* Industry, p. 154.
29. This approach was pioneered in the late nineteenth century by a voluntary organisation, the Mental After Care Association. See V. Long, 'Changing Public Representations of Mental Illness in Britain, 1870–1970' (PhD thesis, University of Warwick, 2004); V. Long, ' "A Satisfactory Job is the Best Psychotherapist": Employment and Mental Health, 1939–60', in P. Dale and J. Melling (eds), *Mental Illness and Learning Disability Since 1850: Finding a Place for Mental Disorder in the United Kingdom* (Abingdon, 2006), pp. 179–199.
30. F. Bramley, 'The Attitude of Labour', *Journal of Industrial Welfare*, 3 (1921), 407–409; 408.
31. J. Dover Wilson, 'Humanism in the Continuation School', *Journal of Industrial Welfare*, 4 (1922), 331–333; 331.
32. Although hours of work were not legally restricted to 48 hours a week for adult women workers until the adoption of the 1937 Factory Act, which laid down no maximum working hours for adult men.
33. The shift in opinion between the First and Second World War regarding the psychological impact of repetitive, monotonous work is explored in further detail in Chapter 1.
34. D. Proud, *Welfare Work: Employers' Experiments for Improving Working Conditions in Factories. With Foreword by David Lloyd George PM and first Minister of Munitions* (3rd edition, London, 1918), p. 97.
35. *Annual Report of the Chief Inspector of Factories for the Year 1933*, (1934), Cmd 4657, pp. 50–51.
36. S. Wyatt and J. N. Langdon with F. G. L. Stock, *Fatigue and Boredom in Repetitive Work: Industrial Health Research Board Report No. 77* (London, 1937).
37. Ibid, p. 10.

38. Gollan, *Youth in British Industry*, p. 190.
39. Modern Records Centre, University of Warwick (hereafter MRC), MSS.292C/65.5/4, 1936 Annual Conferences of Representatives of Trade Unions Catering for Women Workers, typescript.
40. MRC, National Union of Railwaymen Archive, MSS.127/NU/GS/3/87, Amalgamated Engineering Union, 'First Report on Health and Welfare Enquiry: Part II' (1944), quoted on page 86. Typed report printed by the union.
41. B. Muscio, *Vocational Guidance (A Review of the Literature): Reports of the Industrial Fatigue Research Board no. 12* (London, 1921), p. 1.
42. Ibid, pp. 5, 6.
43. F. W. Taylor, *The Principles of Scientific Management* (1911), p. 59 cited in ibid, p. 19.
44. A. S. Otis, 'The Selection of Mill Workers by Mental Tests', *Journal of Applied Psychology*, 4 (1920), 339–341 in ibid, p. 50.
45. Wyatt, Langdon and Stock, *Fatigue and Boredom*, p. 66.
46. 'Editorial Notes', *Industrial Welfare*, 6 (1924), 369–370. It was not uncommon for industrial welfare advocates to criticise scientific management and stress the professional superiority of industrial welfare.
47. C. S. Myers, 'Monotony in Production', *Industrial Welfare*, 7 (1925), 52.
48. Wyatt, Langdon and Stock, *Fatigue and Boredom*, p. 65.
49. Ibid, pp. 67, 71.
50. Ibid, p. 70.
51. This stereotype is discussed in more detail in V. Long and H. Marland, 'From Danger and Motherhood to Health and Beauty: Health Advice for the Factory Girl in Early Twentieth-Century Britain', *Twentieth Century British History*, 20 (2009), 454–481.
52. Long, 'A Satisfactory Job is the Best Psychotherapist'.
53. Wyatt, Langdon and Stock, *Fatigue and Boredom*, p. 72.
54. TUC Archive, MSS.292/140.1/2, J. R. Rees, 'The Problem of Neurotic Disorder in its Relation to Industrial Efficiency', undated.
55. TUC Archive, MSS.292/140.1/2, Dr M. Rosenfield, 'Industrial Mental Hygiene Clinics', *Industrial Welfare and Personnel Management* (February 1935); Memorandum produced by Rosenfield for the TUC, 27 July 1935.
56. TUC Archive, MSS.292/140.1/2, Dr E. N. Snowden, 'Injury Psychoneurosis (Traumatic Neurasthenia)', 1 March 1938.
57. TUC Archive, MSS.292/140.1/2, Amalgamated Engineering Union, document regarding members in mental institutions, May 1938.
58. TUC Archive, MSS.292/140.1/2, J. W. Yerrell, 'Memorandum on Nervous Diseases', October 1937.
59. R. Fraser, *The Incidence of Neurosis Among Factory Workers: Industrial Health Research Board Report No. 90* (London, 1947).
60. Ibid, p. 9.
61. Ibid, p. 41.
62. R. Hyde, 'Editor's Notes', *Boys' and Industrial Welfare Journal*, 1 (1918), 1–4; p. 3.
63. A. Wollacott, 'Maternalism, Professionalism and Industrial Welfare Supervisors in World War 1 Britain', *Women's History Review*, 3 (1994), 29–56; 'The Mother Heart: Welfare and the Underpinning of Domesticity', in D. Thom,

Nice Girls and Rude Girls: Women Workers in World War 1 (2nd edition, London, 2000), pp. 164–186; 'Biology as Destiny: Women, Motherhood and Welfare', in G. Braybon, *Women Workers in the First World War* (London, 1981), pp. 112–153.

64. 'Boy Workers Welfare – The Start at 6AM Too Early', *The Times*, 26 October 1917, p. 2.

65. H. Hendrick, *Images of Youth: Age, Class and the Male Youth Problem, 1880– 1920* (Oxford, 1990).

66. See, for example, B. Harris, 'Educational Reform, National Fitness, and Physical Education in Britain, 1900–1945', and J. Welshman, 'Child Health, National Fitness, and Physical Education in Britain, 1900–1940', both in M. Gijswijt-Hofstra and H. Marland (eds), *Cultures of Child Health in Britain and the Netherlands in the Twentieth Century* (Amsterdam, 2003), pp. 85–102 and pp. 61–84.

67. See D. Thom, 'Healthy Citizen of Empire or Juvenile Delinquent?: Beating and Mental Health in the UK', in Gijswijt-Hofstra and Marland (eds), *Cultures of Child Health in Britain and the Netherlands*, pp. 189–212.

68. Gollan, *Youth in British Industry*.

69. This argument is made at greater length in Long and Marland, 'From Danger and Motherhood'.

70. *Health of Munition Workers Committee Interim Report: Industrial Efficiency and Fatigue* (1917), Cd 8511, p. 103.

71. *Annual Report of the Chief Inspector of Factories and Workshops for the Year 1914* (1915), Cd 8051, p. 52.

72. See the section 'Working Hours and Industrial Fatigue' in Chapter 1 for a discussion of the debate on the impact of long working hours on the health and efficiency of workers in both World Wars.

73. The subject received detailed attention by Hilda Martindale for example in 1926: *Annual Report of the Chief Inspector of the Chief Inspector of Factories and Workshops for the Year 1926* (1927), Cmd 2903, pp. 59–67.

74. *Annual Report of the Chief Inspector of Factories for the Year 1935* (1936), Cmd 5230, p. 42.

75. 'The Sea Cadet Corps: An Opportunity for Shipyard Welfare Workers', *Boys' and Industrial Welfare Journal*, 7 (1919), 99–100; 99.

76. Proud, *Welfare Work*, p. 204.

77. B. Webb, *Health of Working Girls: A Handbook for Welfare Supervisors and Others* (London, 1917), p. 101.

78. These findings were reported in 1929 in a chapter on the subject of weight lifting by Sybil Overton, one of the medical factory inspectors: *Annual Report of the Chief Inspector of Factories and Workshops for the Year 1929* (1930), Cmd 3633, pp. 99–103.

79. This conclusion was drawn in 1926 by Hilda Martindale in a chapter devoted to the subject of the young person in industry: *Annual Report 1926*, pp. 58–67.

80. *Annual Report of the Chief Inspector of Factories for the Year 1944* (1945), Comd 6698. The accident rate per thousand rose from 36 for men aged over 18 and 46 for men aged under 18 in 1939 to a peak of 53 and 59 respectively in 1942 (p. 6). Interestingly, though, male workers aged under 18 suffered far fewer fatal accidents: 0.1 per thousand versus 0.26 for adult men (p. 11).

81. Ibid, p. 11.
82. B. Harrison, *Not Only The 'Dangerous Trades': Women's Work and Health in Britain, 1880–1914* (London, 1996); C. Malone, *Women's Bodies and Dangerous Trades in England, 1880–1914* (Woodbridge, 2003).
83. Wyatt, Langdon and Stock, *Fatigue and Boredom*, p. 10.
84. Fraser, *The Incidence of Neurosis*, p. 9.
85. J. Lewis, *The Politics of Motherhood: Child and Maternal Welfare in England 1900–1939* (London, 1980); D. Dwork, *War is Good for Babies and Other Young Children: A History of the Infant and Child Welfare Movement in England 1898–1918* (London, 1989); A. Davin, 'Imperialism and Motherhood', *History Workshop Journal*, 5 (1978), 9–66.
86. Saving the Child. Mothercraft a Part of Statecraft', *The Times*, 8 May 1916, 10.
87. S. MacDonald, *Simple Health Talks with Women War Workers* (London, 1917).
88. 'Women and War Work II – The Physical Recreation', *The Times*, 4 October 1916, 11.
89. This is discussed in Chapter 1.
90. MacDonald, *Simple Health Talks*, p. 8.
91. Mass Observation, *War Factory: A Report* (London, 1943), pp. 85–86.
92. See Thom, *Nice Girls*, pp. 174–178; Braybon, *Women Workers*, 112–153; P. Summerfield, 'Women, Work and Welfare: A Study of Child Care and Shopping in Britain in the Second World War', *Journal of Social History*, 17 (1983), 249–269.
93. *The Annual Report of The Chief Inspector of Factories and Workshops for the Year 1918* (1919), Cmd 340, pp. 52–56; p. 53.
94. Proud, *Welfare Work*, p. 110. These trends are discussed further in Long, 'Industrial Homes'.
95. S.W., 'The Problem of Home and Factory', *Journal of Industrial Welfare*, 4 (1922), 25–26.
96. *Annual Report of the Chief Inspector of Factories for the Year 1933*, pp. 50–51.
97. Ibid, p. 51.
98. TUC Archive, MSS.292C/65.5/4, 1936 Annual Conferences of Representatives of Trade Unions Catering for Women Workers, typescript.
99. TUC Archive, MSS.292/134.1/5. The total approved membership was given as 395,373 men and 195,491 women, with an average of 146,630 men and 42,114 women drawing sickness and disablement benefit per year for the past five years, equating to 37 per cent of the men and 21.5 per cent of women who were members of the approved societies.
100. More broadly, Susan Aspinall's doctoral research indicates that from around 1900 onwards, biological determinism co-existed and competed with environmentalist explanations for women's health: S. Aspinall, 'Nature as well as Nurture: Environmentalism in Representations of Women and Exercise in Britain from the 1880s to the Early 1920s' (PhD thesis, University of Warwick, 2009).
101. S. Sturdy and R. Cooter, 'Science, Scientific Management, and the Transformation of Medicine in Britain C. 1870–1950', *History of Science*, 36 (1998), 1–47.

102. Sturdy elaborates on this point at greater length in S. Sturdy, 'Hippocrates and State Medicine: George Newman Outlines the Founding Policy of the Ministry of Health', in C. Lawrence and G. Weisz (eds), *Greater than the Parts: Holism in Biomedicine 1920–1950* (Oxford, 1998), pp. 112–134. Nigel Oswald has argued that the failure of this campaign helps explain why the introduction of the NHS failed to tackle preventable and chronic illness; in short, while medical practice was transformed, training was not: N. Oswald, 'Training Doctors for the National Health Service: Social Medicine, Medical Education and the GMC 1936–48', in D. Porter (ed.), *Social Medicine and Medical Sociology in the Twentieth Century* (Amsterdam, 1997), pp. 59–80.
103. J. A. C. Brown, *Social Psychology in Industry* (1954), p. 130, cited in Rose, *Governing the Soul*, p. 85.
104. Ministry of Health, *A National Health Service* (1944), Cmd 6502, p. 10.

5 A National Industrial Health Service?

1. Modern Records Centre, University of Warwick (hereafter MRC), National Union of Railwaymen Archive, MSS.127/NU/GS/3/87, Amalgamated Engineering Union, 'First Report on Health and Welfare Parts I and II', (1944), typescript reports.
2. See C. Wrightman, *More Than Munitions: Women, Work and the Engineering Industries 1900–1950* (Harlow, 1999), pp. 133–155.
3. 'First Report on Health and Welfare', p. 9.
4. National Union of Railwaymen Archive, MSS.127/NU/GS/3/87, cutting from the *Amalgamated Engineering Union Journal*, February (1946).
5. Factory Department: Ministry of Labour and National Service, *Memorandum on Medical Supervision in Factories Form 327* (London, 1946).
6. BMA, *Committee on Industrial Health in Factories* (1941).
7. MRC, TUC Archive, MSS.292/142/1, Association of Industrial Medical Officers, 'The Place of an Industrial Medical Service in a Post-War Comprehensive National Health Service' (1943).
8. TUC Archive, MSS.292/142/1, Association of Scientific Workers, 'Memorandum on the Industrial Medical Services' (1944).
9. See J. T. Carter, 'Fifty Years of Medicine in the Workplace: The Contribution of the Society of Occupational Medicine to Occupational Health Practice, 1935–1985', *Journal of the Society of Occupational Medicine*, 35 (1985), 4–22.
10. Association of Scientific Workers, 'Memorandum on the Industrial Medical Services'.
11. TUC Archive, MSS.292/142.1, 1944 Blackpool Congress Resolution, 'Industrial Medical Service'.
12. Ministry of Health, *A National Health Service* (1944), Cmd 6502.
13. Ibid, p. 10.
14. These issues came to the fore in the proceedings of the Dale Committee, published in 1951 and explored in detail in the next chapter: *Report of a Committee of Enquiry on Industrial Health Services* (1951) Cmd 8170.
15. *A National Health Service*, p. 10.

16. A detailed account of the views expressed by different medical, professional and political organisations leading up to the establishment of the NHS is given in C. Webster, *The Health Services Since the War Volume I: Problems of Health Care the National Health Service Before 1957* (London, 1988), pp. 44–94.

17. *Hansard*, 16 March 1944, vol. 398, col 493.

18. Ibid, 493–497.

19. See Webster, *The Health Services Since the War Volume I*, pp. 59, 82, 89.

20. National Archives, Public Records Office (hereafter PRO), MH 80/30, 'Notes on main points taken in BMA Council's Report on Bill', February 1946.

21. Select Committee on National Expenditure, *Health and Welfare of Women in War Factories* (London, 1942), discussed in Chapter 1.

22. J. T. Carter noted that Brunner Mond, one of three companies which merged to form Imperial Chemical Industries in 1926, employed salaried doctors prior to 1914: Carter, 'Fifty Years of Medicine', 5. See also R. Fitzgerald, *British Labour Management and Industrial Welfare, 1846–1939* (London, 1988), pp. 118–125. However, David Walker's research demonstrated that employers in the British Chemical Industry, including Imperial Chemical Industries, prioritised profit over workers' health. Health and welfare measures were designed to promote productivity and to minimise, but not remove, risk: see D. Walker, 'Occupational Health and Safety in the British Chemical Industry, 1914–1974' (PhD Thesis, University of Strathclyde, 2007).

23. TUC Archive, MSS.292/141.2/5, minutes of fourth meeting of the IHAC, 9 November 1944.

24. Ibid.

25. TUC Archive, MSS.292/142/1, TUC deputation to the Minister of Labour and National Service, 16 May 1945.

26. TUC Archive, MSS.292/142/1, TUC Social Insurance Committee, 'Summary of Replies to Questionnaire Issue by Sub-Committee on Development of Industrial Medical Services in Factories', 10 November 1948.

27. Isaacs served as General Secretary of the National Society of Operative Printers and Assistants and Chairman of the TUC General Council in 1945. He was appointed Minister of Labour and National Service by Clement Attlee in 1945, replacing the previous incumbent Ernest Bevin. See H. Pemberton, 'Isaacs, George Alfred (1883–1979)', *Oxford Dictionary of National Biography* (Oxford, 2004), www.oxforddnb.com/view/article/32175, accessed 7 June 2010.

28. TUC Archive, MSS.292/142/1, TUC memorandum, 'Industrial Medial Services (Factories) Notes for Deputation to Ministry of Labour on 27 July 1948'.

29. Ibid.

30. TUC Archive, MSS.292/142/1, *Financial Times*, 8 July 1948.

31. TUC Archive, MSS.292/142/1, 'Notes of a Deputation from the TUC General Council Deputation received by the Ministry of Labour on 27 July 1948'.

32. PRO, PREM 8/1487, letter from George Isaacs to Clement Attlee and Aneurin Bevan, 6 August 1948.

33. PRO, LAB 14/689, note to Mr Rossetti, 8 January 1949.

34. *Report of a Committee of Enquiry on Industrial Health Services*.

35. *Health, Welfare and Safety in Non-Industrial Employment* (1949), Cmd 7664; this Committee and its findings are discussed in more detail in the following chapter.
36. PRO, PREM 8/1487, letter from Aneurin Bevan to Clement Attlee, 18 December 1950.
37. PRO, PREM 8/1487, letter from George Isaacs to Clement Attlee, 20 December 1950.
38. PRO, PREM 8/1487, letter from Aneurin Bevan to Clement Attlee, 11 January 1951.
39. PRO, LAB 14/712, note from Mary Smieton to Miss Fox, 19 January 1954.
40. 'Industrial Health Services', *The Lancet*, 257:6654 (10 March 1951), 567–568; 568.
41. See M. Pugh, 'Monckton, Walter Turner, first Viscount Monckton of Benchley (1891–1965)', *Oxford Dictionary of National Biography* (Oxford, 2004), www.oxforddnb.com/view/article/35061, accessed 10 June 2010.
42. TUC Archive, MSS.292/142.9/2, National Joint Advisory Council Paper 139 (1952), 'The Dale Committee's Report on Industrial Health Services'.
43. Ibid.
44. PRO, LAB 14/712, note to Mr Buckland, 16 May 1952.
45. TUC Archive, MSS.292C/140/1, TUC Social Insurance and Industrial Welfare Committee, 'Occupational Health Service', 8 October 1952.
46. TUC Archive, MSS.292/142.9/2, letter from Vincent Tewson to Walter Monckton, 2 July 1952; letter from Monckton to Tewson, 16 July 1952.
47. TUC Archive, MSS.292/142.9/2, notes for Mr Roberts (probably Mr A. Roberts, National Association of Card, Blowing and Ring Room Operatives) on the IHAC and Dale Committee Report on Industrial Health Services, undated circa 1952.
48. MSS.292C/140/1, letter from Walter Monckton to Vincent Tewson, 5 November 1952: emphasis in original letter.
49. PRO, LAB 14/712, draft letter to Sir Vincent Tewson and accompanying notes, 24 October 1952.
50. TUC Archive, MSS.292/142.9/2, TUC, 'Comprehensive Occupational Health Service: Notes for Deputation to Minister of Labour', 24 February 1953.
51. Ibid.
52. PRO, LAB 14/712, note from H. R. Hodges to Dame Mary Smieton, 13 August 1953.
53. These proposals were drawn up largely by Dame Mary Smieton and Sir Godfrey Ince, respectively the Under Secretary and Secretary for the Ministry of Labour and Sir Guildhaume Myrddin-Evans, head of the international division of the Ministry of Labour, See G. Cockerill, 'Smieton, Dame Mary Guillan (1902–2005)', *Oxford Dictionary of National Biography* (Oxford, 2009), http://www.oxforddnb.com/view/article/65554, J. Saville, 'Evans, Sir Guildhaume Myrddin- (1894–1964), rev. *Oxford Dictionary of National Biography* (Oxford, 2004), http://www.oxforddnnb.com/view/article/35179 and R. Lowe, 'Ince, Sir Godfrey Herbert (1891–1960)', *Oxford Dictionary of National Biography* (Oxford, 2004), http://www.oxforddnb.com/view/article/34095: all consulted 10 June 2010.
54. PRO, LAB 14/712, note from Dame Mary Smieton to Sir Guildhaume Myrddin-Evans, 7 November 1953.

55. PRO, LAB 14/712, Mary Smieton, 'An Occupational Health Service', 6 October 1953.
56. PRO, LAB 14/712, letter from Sir Walter Monckton to Viscount Woolton, 31 December 1953.
57. PRO, LAB 14/712, letter from Vincent Tewson to Walter Monckton, 24 February 1954; notes of deputation received by the Prime Minister from the TUC, 25 February 1954.
58. PRO, LAB 14/712, 'Note of meeting with the BMA', 22 January 1954; 'Note of meeting with the BMA', 16 March 1954.
59. PRO, CAB 129/65, Minister of Labour and National Service, 'A National Occupational Health Scheme', 7 January 1954 C (54) 8 and PRO, CAB 129/66, Minister of Health, 'A National Occupational Health Scheme', 11 March 1954 C (54) 98. Memoranda discussed at a meeting of the Cabinet, 28 April 1954: PRO, CAB/128/27. See also the minutes of the meeting: PRO, CAB/195/12.
60. PRO, LAB 14/712, 'National Occupational Health Scheme', undated, c. March/April 1954.
61. PRO, CAB/128/27, conclusions of a meeting of Cabinet, 28 July 1954.
62. For a detailed account of these developments, see N. Tiratsoo and J. Tomlinson, *Industrial Efficiency and State Intervention: Labour 1939–51* (London, 1993).
63. MRC, BEC Archive, MSS.200/B/3/2/C1024 pt 2, letter from John Sadd to Kenneth Burton, 'Proposed Industrial Health Services', 4 October 1954.
64. BEC Archive, MSS.200/B/3/2/C1024 pt 2, letter from George Pollock, Director of the Confederation, to John Sadd, 6 October 1954.
65. PRO, LAB 14/638, note, signed Iain MacLeod, 26 March 1956.
66. *Industrial Health: A Survey of Halifax. Report by Her Majesty's Factory Inspectorate and Recommendations of the IHAC* (London, 1958).
67. PRO, LAB 14/1010, report of a sub-committee of the IHAC, 'Pilot Industrial Health Survey – Halifax', 1956, p. 22, p. 37, p. 1.
68. A. Meiklejohn, 'Halifax, Stoke-on-Trent – Whither?', *Transactions of the Society of Occupational Medicine*, 9 (1960), 143–144; 143.
69. PRO, LAB 14/806, 'Statement from the Halifax Industrial Health Committee', date stamped 23 June 1961.
70. Ibid, letter from J. W. McAnuff, Assistant Director of the Nuffield Foundation to the joint secretaries of the Halifax Industrial Health Committee, 21 February 1963.
71. *Industrial Health: A Survey of the Pottery Industry in Stoke-on-Trent* (London, 1959).
72. Meiklejohn, 'Halifax, Stoke-on-Trent', 144.
73. *Hansard*, 29 January 1957, vol. 563, cols 837–838.
74. TUC Archive, MSS.292C/140/2, TUC Social Insurance and Industrial Welfare Department, 'Industrial Health Services', September 1957.
75. PRO, LAB 14/1515, 'Appendix II: Comparison between the position in February, 1965 and that at February, 1963'.
76. TUC Archive, MSS.292/142.9/2, conveyed in letter from V. Tewson to E. Heath, 1 July 1960.
77. International Labour Office, *Report IV (2) of the Forty-Third Session of the International Labour Conference* (1959), cited in 'Occupational Health Services', *British Medical Journal*, 2:5243 (1 July 1961), 37–38.

78. PRO, LAB 14/638, note from Mr Rosetti to Mr Barnes, 9 November 1959.
79. TUC Archive, MSS.292/142.9/2, conveyed in letter from Tewson to Heath, 1 July 1960.
80. PRO, LAB 14/638, 'Memorandum on Industrial Health', prepared for meetings, 23–24 January 1960.
81. TUC Archive, MSS. 292/142.9/2, letter from Heath to Tewson, 27 July 1960, enclosing extract from Hansard containing a statement made by Heath on 25 July 1960.
82. Ibid.
83. PRO, LAB 14/1154, note from Mr Singleton to Mr Rosetti, 19 November 1962. The Porritt Committee contained representatives from the Royal College of Physicians, the Royal College of Surgeons, the Royal College of Obstetricians and Gynaecologists, the Society of Medical Officers of Health, the College of General Practitioners and the British Medical Association.
84. Medical Services Review Committee, *A Review of the Medical Services in Great Britain* (London, 1962); British Medical Association, *The Future of Occupational Health Services* (London, 1961).
85. D. Stewart, 'The Future of Occupational Health', *British Medical Journal*, 1:4646 (21 January 1950), 156–159; 'Occupational Health Service', *British Medical Journal*, 2:4898 (20 November 1954); 1214–1216.
86. 'Occupational Health Services', *British Medical Journal*, 2:5243 (1 July 1961), 37–38.
87. TUC Archive, MSS.292/142/1, minutes from Workmen's Compensation and Factories Committee, 7 May 1947.
88. TUC Archive, MSS. 292/142/1, 'A State Experiment in Industrial Hygiene Service for Small Plants', notes sent by Dr Schilling to C. R. Dale, 20 July 1947.
89. The emergence of the occupational health team and the concurrent displacement of the figure of the industrial medical officer has been examined in Carter, 'Fifty Years of Medicine'.
90. TUC Archive, MSS.292/142/32/2, 'Slough Industrial Health Service'; July 1956. By 1962 the scheme covered 275 factories employing 26,000 workers.
91. TUC Archive, MSS.292/142/32/2, memorandum of interview between Mr Owen of the Social Insurance Department and Mr L. Leyshon and Mr F. C. Evennett of the Slough Trades Council, 20 October 1947.
92. TUC Archive, MSS.292/142.9/2, TUC Social Insurance and Industrial Welfare Committee memorandum, 'Industrial Health: Propose Scheme for Provision of an Industrial Health Service for Grouped Undertakings', 11 June 1952.
93. Ibid.
94. TUC Archive, MSS.292/142/5, TUC Report, 'Existing Group Industrial Health Services', 17 December 1962.
95. R. Johnston and A. McIvor, 'Whatever Happened to the *Occupational* Health Service? The NHS, the OHS and the Asbestos Tragedy on Clydeside', in C. Nottingham (ed.), *The NHS in Scotland: The Legacy of the Past and the Prospect of the Future* (Aldershot, 2000), pp. 79–105; 90.
96. F. H. Tyrer, 'Group Occupational Health Schemes', *Journal of the Society of Occupational Medicine*, 30 (1980), 118–122. Tyrer served as President of the

Society of Occupational Medicine in 1980: see R. M. Archibald, 'Obituary: Frank H. Tyrer', *Journal of the Society of Occupational Medicine*, 36 (1986), 73.

97. Tyrer, 'Group Occupational Health Schemes', 118.
98. Ibid.
99. *Hansard*, 11 May 1964, vol 695, cols. 25–27.
100. Tyrer, 'Group Occupational Health Services', 122.
101. These services are members of the Association of Grouped Occupational Health Services; further information can be found at www.agohs.org.uk/index.htm, accessed 31 July 2008.
102. MSS.292C/140/2, TUC Social Insurance and Industrial Welfare Department, 'Industrial Health Services', September 1957.
103. One recent correspondent to *Occupational Medicine* recalled: 'Fifty years ago "Industrial Medicine" really was a Cinderella among medical specialities, and doctors often entered it "faute de mieux" when they could not obtain a specialist or hospital appointment.' H. Engel, 'Why I Became an Occupational Physician', *Occupational Medicine*, 58 (2009), 151.
104. Leading article, 'Progress – and Retrogression', *Transactions of the Association of Industrial Medical Officers*, 9 (1960), 115.
105. *Hansard*, 6 May 1980, vol. 984 col. 5: by this stage the Home Office Industrial Museum was operating as the Industrial Health and Safety Centre.
106. In one sense this reflects the thesis advanced by Keith Middlemas and discussed in Chapter 3 that the State prevented political upheaval by incorporating employer associations and trade unions into governance in the years 1916 to 1962: See K. Middlemas, *Politics in Industrial Society: The Experience of the British System Since 1911* (London, 1979). More generally, historians have argued that both main political parties pursued a similar set of collectivist policies in the post-war years until economic problems in the 1970s forced the parties apart. An overview of this debate is provided in S. Glynn and A. Booth, *Modern Britain: An Economic and Social History* (London, 1996), pp. 184–187.
107. Tiratsoo and Tomlinson, *Industrial Efficiency*, p. 169.
108. M. Greenberg, 'The Last Senior Medical Inspector of Factories and His Place in the History of Occupational Health', *American Journal of Industrial Medicine*, 49 (2006), 54–59; 54.
109. See G. R. Searle, *The Quest for National Efficiency* (Oxford, 1971); S. Sturdy, 'The Industrial Body', in R. Cooter and J. Pickstone (eds), *Companion to Medicine in the Twentieth Century* (London, 2003), pp. 213–234.

6   The Fall of the Healthy Factory

1. Modern Records Centre, University of Warwick (hereafter MRC), TUC Archive, MSS.292/140/3, pamphlet for conference, 'The Role of Industrial Medicine in the Welfare State'.
2. *Annual Report of the Chief Inspector of Factories for the Year 1943* (1944), Cmd 6563, p. 41.
3. TUC Archive, MSS.292/142/1, 'Industrial Medical Service (Factories)', statement of government policy sent by the Ministry of Labour and National

Service to Vincent Tewson, 30 August 1948. A TUC deputation to George Isaacs, the Minister of Labour, prompted this statement of policy: the deputation is discussed in Chapter 5.

4. J. T. Carter, 'Fifty Years of Medicine in the Workplace: The Contribution of the Society of Occupational Medicine to Health Practice 1935–85', *Journal of the Society of Occupational Medicine*, 35 (1985), 4–22; 12.

5. TUC efforts in this field were summarised in a memorandum prepared by Dr Morgan for Walter Citrine: TUC Archive, MSS.292/140/2 TUC, 'Industrial Medicine', 5 October 1944.

6. Ministry of Health, *A National Health Service* (1944), Cmd 6502.

7. National Archives, Public Records Office (PRO), PREM 8/1487, minutes from a meeting of the Socialised Industries Committee, 1 December 1948.

8. PRO, PREM 8/1487, joint memorandum by the Minister of Health and the Secretary of State for Scotland, 'Socialisation of Industries Committee: Co-ordination of Health Services' S.I. (M) (49) 5, 12 January 1949.

9. PRO, PREM 8/1487, letter from George Isaacs to H. S. Morrison, Lord President of the Council, 13 January 1949.

10. PRO, PREM 8/1487, minutes of the Socialisation of Industries Committee, 25 January 1949; joint memorandum by the Lord President of the Council and the Chancellor of the Exchequer, 3 March 1949.

11. *Hansard*, 1 June 1949, volume 465, cols 2118–2120.

12. Ibid.

13. MRC, BEC Archive, MSS.202/B/2/C1024 pt 1, Ministry of Health, 'Industrial Health Services Committee', memorandum of evidence.

14. Ibid.

15. BEC Archive, MSS.202/B/2/C1024 pt 1, General Baylay; Dr Caldwell-Smith: minutes of meeting of Factories Standing Committee, 12 October 1949.

16. BEC Archive, MSS.202/B/2/C1024 pt 1, this opinion was ventured by a member of the Coventry and District Engineering Employers' Association in a set of papers dispatched to the BEC: 'Industrial Health Services – Government Committee', 27 September 1949.

17. BEC Archive, MSS.202/B/2/C1024 pt 1, letter from the National Federation of Vehicle Trades to the BEC, 30 September 1949.

18. BEC Archive, MSS.202/B/2/C1024 pt 1, 'Industrial Health Services Committee: Meeting held at the offices of the Industrial Welfare Society on February 22 1950', transcript of oral evidence given by the BEC to the IHSC.

19. See R. Dingwell, A. M. Rafferty and C. Webster, *An Introduction to the Social History of Nursing* (London, 1988) pp. 98–105; 116–118.

20. TUC Archive, MSS.292/142/1, interdepartmental TUC memorandum from E. P. Harries to C. R. Dale, 'Medical Occupational Committee of the BMA', 23 July 1948.

21. In 1943, the Factory Inspectorate noted that 8,395 nurses were employed in factories, of which around half were state registered: *Annual Report of the Chief Inspector of Factories for the Year 1943*, p. 53. The 1947 figures are those quoted in the report issued by the IHSC: *Report of a Committee of Enquiry on Industrial Health Services* (1951), Cmd 8170, p. 7.

22. Nevil Johnson, 'Vickers, Sir (Charles) Geoffrey (1894–1982)', rev., *Oxford Dictionary of National Biography* (Oxford, 2004), www.oxforddnb.com/view/

article/31787, accessed 25 June 2008. Given Vickers' approach it is worth noting that the National Coal Board had a well-developed industrial medical service in place, although did little to bring levels of dust down below approved levels: see A. McIvor and R. Johnston, *Miners' Lung: A History of Dust Disease in British Coal Mining* (Aldershot, 2007), pp. 105–111; 145–183.

23. BEC Archive, MSS.202/B/2/C1024 pt 1, oral evidence given by the Ministry of Health to the IHSC, 23 September 1949, typescript.

24. Ibid.

25. And indeed, this was the number given in the Ministry of Health's memorandum for the IHSC: BEC Archive, MSS.202/B/2/C1024 pt 1, Ministry of Health, 'Industrial Health Services Committee'.

26. TUC Archive, MSS.292/142.9/2, memorandum submitted to the (BMA?) Committee on Nursing by the (BMA?) Occupational Health Committee, 'The Economic Use of Nurses in Industry', 14 September 1949. BMA memoranda probably found their way into TUC files because of Dr Morgan's involvement in both organisations.

27. BEC Archive, MSS.202/B/2/C1024 pt 1, IHSC: minutes of meeting held 25 August 1949 MIN/3/49. Discrepancies in the figures given by different organisations arose because some organisations included the number of doctors employed as Appointed Factory Doctors under the terms of the Factories Act; as such doctors often undertook additional duties in excess of the minimum legal requirements it can be difficult to distinguish the extent of voluntary industrial health service provision from the meeting of statutory legal requirements.

28. TUC Archive, MSS. 292/147.6/2, Ministry of Labour and National Service Press Notice, 'Factory Medical Services', 18 May 1949. This survey was undertaken at the request of the Sub-Committee on the Development of Industrial Medical Services in Factories of the Industrial Health Advisory Committee, discussed in Chapter 5: See National Archives, PRO, LAB 14/690-91.

29. BEC Archive, MSS.202/B/2/C1024 pt 1, oral evidence given by the Ministry of Health.

30. Ibid.

31. Ibid.

32. National Archives, PRO, LAB 14/496, note from Guildhaume Myrddin-Evans to Godfrey Ince, 2 September 1949.

33. PRO, LAB 14/496, note from E. R. A. Merewether to Mr Buckland, 4 October 1949.

34. PRO, LAB 33/2, IHSC: oral evidence given by Sir Godfrey Ince and Mr G. R. Buckland on behalf of the Ministry of Labour and National Service, 5 October 1949.

35. BEC Archive, MSS.202/B/C1024 pt 1, Ministry of Labour and National Service, 'Industrial Health Services Committee: Factories with Special Industrial Hazards', 1949.

36. BEC Archive, MSS.202/B/2/C1024, transcript of oral evidence given by the BEC to the IHSC.

37. BEC Archive, MSS.202/B/2/C1024 pt 1, oral evidence given by the Ministry of Health.

38. Ibid.

39. BEC Archive, MSS.202/B/2/C1024 pt 1, Coventry and District Engineering Employers' Association, 'Industrial Health Services – Government Committee'.
40. Ibid.
41. BEC Archive, MSS.202/B/2/C1024 pt 1, oral evidence given by the Ministry of Health to the IHSC.
42. BEC Archive, MSS.202/B/2/C1024 pt 1, Coventry and District Engineering Employers' Association, 'Industrial Health Services – Government Committee'.
43. TUC Archive, MSS.292/142.9/2, transcript of TUC deputation to the IHSC, 25 January 1950.
44. C. R. Dale, 'Viewpoints on an Occupational Health Service', *Transactions of the Association of Industrial Medical Officers*, 2 (1953), 164–167; 164.
45. BEC Archive, MSS.202/B/2/C1024 pt 2, minutes of meeting of the BEC's "Factories" Standing Committee, held on 12 October 1949.
46. TUC Archive, MSS.292/142/1, TUC deputation to Ministry of Labour and National Service, 16 May 1945.
47. PRO, LAB 33/2. This position was adopted, for example, by the Royal College of Physicians and the Ministry of National Insurance. See also evidence given by the Society of Medical Officers of Health: PRO, LAB 33/4.
48. Home Office and Scottish Home Department, *Health, Welfare, and Safety in Non-Industrial Employment: Hours of Employment of Juveniles* (1949), Cmd 7664. The Committee was also instructed to investigate current provisions of the Shops Acts as regards closing hours and to inquire into the statutory regulation of hours of employment of young persons.
49. Ibid, p. 12.
50. Ibid, pp. 7–8.
51. *Annual Report of the Chief Inspector of Factories and Workshops for the Year 1913* (1914), Cd 7491, p. 65.
52. *Reports of the Proceedings of the Forty-Sixth Annual Trades Union Congress* (London, 1913), p. 215.
53. TUC Archive, MSS.292C/65.5/1/5, *Report of the Eighth Annual Conference of Unions Catering for Women Workers, April 23 1938* (London), p. 15.
54. TUC Archive, MSS.292C/65.5/1/12, *Report of the Fifteenth Annual Conference of Unions Catering for Women Workers, October 12–13 1945* (London), p. 29.
55. A. Marshall, *Principles of Economics* (1890), footnote 124; Oxford English Dictionary, (2nd edition, 1989), consulted online at http://dictionary.oed.com/ on 4 February 2008.
56. TUC Archive, MSS.292C/65.5/1/17, *Report of the Twentieth Annual Conference of Representatives of Trade Unions Catering for Women Workers, May 12–13, 1950* (London), p. 6.
57. Ibid, p. 29.
58. Jean Bunce has explored how greater use of machinery and chemicals within agriculture increased the range of occupational accidents and diseases to which British farmers were exposed in the twentieth century. Her study also considered the specific occupational factors which produce stress and mental health problems amongst agricultural workers. See J. Bunce, 'The "Healthiest of Occupations" or One of the Most Dangerous? The Health Hazards of Agriculture in Britain 1934–2004' (MA thesis, University of

Warwick, 2004). Although not explored within this monograph, the ways in which concern with industrial disease evolved in America into environmental health science, a discipline which explored the impact of industrial production on the environment more generally, formed the focus for Chris Sellers' 1997 monograph: C. Sellers, *Hazards of the Job: From Industrial Disease to Environmental Health Science* (Chapel Hill, 1997).

59. The findings of this Committee were published in 1951: *Report of a Committee of Enquiry on Industrial Health Services*.

60. TUC Archive, MSS.292/142/1, 'Industrial Medical Services (Factories)': Statement of government policy sent by the Ministry of Labour and National Service to Vincent Tewson, 30 August 1948.

61. TUC Archive, MSS.292/142/1, 'Medical Occupational Committee of the BMA', interdepartmental memorandum from E. P. Harries to C. R. Dale, 23 July 1948. Sir George Schuster chaired the panel on human factors, an offshoot of the Committee on Industrial Productivity established in 1947. See N. Tiratsoo and J. Tomlinson, *Industrial Efficiency and State Intervention: Labour 1939–51* (London, 1993), pp. 92–93.

62. See C. Gordon, *Dead on Arrival: The Politics of Health Care in Twentieth-Century America* (Princeton, 2003).

63. R. Johnston and A. McIvor, 'Whatever Happened to the *Occupational* Health Service? The NHS, the OHS and the asbestos tragedy on Clydeside', in C. Nottingham (ed.), *The NHS in Scotland: The Legacy of the Past and the Prospect of the Future* (Aldershot, 2000), pp. 79–105; R. Johnston and A. McIvor, 'Marginalising the Body at Work? Employers' Occupational Health Strategies and Occupational Medicine in Scotland c. 1930–1974', *Social History of Medicine*, 21 (2008), 127–144.

64. C. Webster, *The Health Services Since the War Volume 1: Problems of Health Care – The National Health Service Before 1957* (London, 1988), p. 133.

65. Statistics from ibid, Table 1: The Cost of the National Health Service 1948–1957, p. 135. The efforts of successive governments to cut back on NHS expenditure is discussed in ibid, pp. 133–215.

66. Professor Lee, employed within the Occupational Health Department at Manchester University before his retirement. These comments were made at a seminar on 4 March 2008.

67. PRO, LAB 14/496, note from Mr Sisson to Mr Hodges, undated, circa. September 1949.

68. *Safety and Health at Work* (1972), Cmnd 5034.

69. Ibid, p. 115.

Conclusion

1. C. Black, *Working for a Healthier Tomorrow: Review of the Health of Britain's Working Age Population* (London, 2008), p. 12: The term 'health conditions' is undoubtedly used by Black to facilitate her objective of tackling the stigma surrounding the employment of people who experience 'health problems'.

2. P. W. J. Bartrip and S. B. Burman, *The Wounded Soldiers of Industry: Industrial Compensation Policy 1833–1897* (Oxford, 1983), p. 211.

3. Ibid, p. 211; in 1906 a number of scheduled industrial diseases were brought within the ambit of the Act for the first time.

4. Ibid, p. 211.

5. A. McIvor and R. Johnston, *Miner's Lung: A History of Dust Disease in British Coal Mining* (Aldershot, 2007), pp. 123–144.

6. E. Scarry, *The Body in Pain: The Making and Unmaking of the World* (Oxford, 1985), p. 3.

7. Ibid, p. 5.

8. J. Greenlees, '"Stop Kissing and Steaming!" Tuberculosis and the Occupational Health Movement in Massachusetts and Lancashire, 1870–1918', *Urban History*, 32 (2005), 223–246.

9. The Scottish TUC, established in 1897, worked independently of the TUC; it was not possible within the scope of this project to examine its archives.

10. Membership and density statistics until 1945 from G. Bain and R. Price, *Profiles of Union Growth* (Oxford, 1980), pp. 37–38, cited in A. McIvor, *A History of Work in Britain, 1880–1950* (Basingstoke, 2001), p. 201; for the post-1945 period, S. Glynn and A. Booth, *Modern Britain: An Economic and Social History* (London, 1996), p. 295. These national aggregates disguise differential rates of trade union membership among men and women, in different regions and different occupations.

11. McIvor and Johnston, *Miners' Lung*; R. Johnston and A. McIvor, *Lethal Work: A History of the Asbestos Tragedy in Scotland* (East Linton, 2000); N. Hayes, 'Did Manual Workers Want Industrial Welfare? Canteens, Latrines and Masculinity on British Building Sites', *Journal of Social History*, 35 (2002), 637–658.

12. P. Weindling, 'Linking Self Help and Medical Science: The Social History of Occupational Health', in P. Weindling (ed.), *The Social History of Occupational Health* (London, 1985), pp. 2–31; G. Tweedale, *Magic Mineral to Killer Dust: Turner and Newall and the Asbestos Hazard* (Oxford, 2000).

13. S. Bowden and G. Tweedale, 'Mondays without Dread: The Trade Unions Response to Byssinosis in the Lancashire Cotton Industry in the Twentieth Century', *Social History of Medicine*, 16 (2003), 79–96; McIvor and Johnston, *Miners' Lung*.

14. J. Melling, 'The Risks of Working and the Risks of Not Working: Trade Unions, Employers and Responses to the Risk of Occupational Illness in British Industry, c. 1890–1940s', *ESRC Centre for Analysis of Risk and Regulation Discussion Paper 12* (2003), pp. 14–34; p. 16.

15. While acknowledging the difficulty of distinguishing in practice between measures designed to promote health and measures implemented to treat or compensate injured and sick workers, I have consciously excluded records appertaining to attempts to identify industrial toxins, to introduce regulations to minimise risk or to secure compensation for industrial diseases. The TUC was very active in these field, as can bee seen in the files it generated on Workmen's Compensation (MRC: MSS.292/143) industrial disease and injuries (MRC: MSS.292/144) and accidents (MRC: MSS.292.146).

16. This survey did not attempt to assess BEC initiatives in the field of industrial illness; likewise, interesting though it may have proved to be, it was beyond the remit of this study to examine what work if any was done by different employers' associations in the field of industrial health.

17. For example, Modern Records Centre, University of Warwick (MRC), BEC Archive, MSS.202/B/2/C1024 pts 1–2, correspondence, reports and memoranda; discussed in Chapter 6.

18. See BEC Archive, MSS.200/B/3/2/C693 pts 1–5; discussed in Chapter 3.

19. M. W. Bufton and J. Melling, ' "A Mere Matter of Rock": Organised Labour, Scientific Evidence and British Government Schemes for Compensation of Silicosis and Pneumoconiosis amongst Coalminers, 1926–1940', *Medical History*, 49 (2005), 155–178; 156–157.

20. K. Middlemas, *Politics in Industrial Society: The Experience of the British System Since 1911* (London, 1979).

21. D. Edgerton, *Warfare State: Britain, 1920–1970* (Cambridge, 2006).

22. Exceptions being Gillian Darley's history of factory architecture: G. Darley, *Factory* (London, 2003); Joan Skinner's work on the factories produced by the architectural firm Wallis, Gilbert and Partners: J. Skinner, *Form and Fancy: Factories and Factory Buildings by Wallis, Gilbert and Partners, 1919–1939* (Liverpool, 1997); Michael Stratton and Barrie Trinder's archaeological study of industrial buildings: M. Stratton and B. Trinder, *Twentieth-Century Industrial Archaeology* (London, 2000) and Helena Chance's research on factory gardens in Britain and America: H. Chance, 'The Angel in the Garden Suburb: Arcadian Allegory in the "Girls' Grounds" at the Cadbury Factory, Bournville, England, 1880–1930', *Studies in the History of Gardens and Designed Landscapes*, 27 (2007), 197–216.

23. J. F. C. Harrison, *Robert Owen and the Owenites in Britain and America: The Quest for the New Moral World* (2nd edition, Aldershot, 1994).

24. These cases were detailed in the monthly 'Legal and Statistical Notes' column of the *Journal of Industrial Welfare* in the 1920s; for example, 'When does Employment Cease?', 7 (1925), 281 and 'Accident on the Way to Work – When does Employment Commence?', 8 (1926), 273.

25. MRC, TUC Archive, MSS.292C/146.9/6, TUC General Council, 'The Effect of Welfare Work on Trade Unionism', 24 May 1932.

26. *Report of a Committee of Enquiry on Industrial Health Services* (1951), Cmd 8170.

27. For an overview of these developments see J. T. Carter, 'Twenty-One Years of EMAS', *Occupational Medicine*, 44 (1994), 119–122.

28. *Safety and Health at Work* (1972), Cmnd 5034.

29. See R. Squire, *Thirty Years in the Public Service: An Industrial Retrospect* (London, 1927), p. 45: discussed in Chapter 2.

30. The implementation of the 1974 Act is discussed in R. Johnston and A. McIvor, 'Whatever Happened to the *Occupational* Health Service? The NHS, the OHS and the Asbestos Tragedy on Clydeside', in C. Nottingham (ed.), *The NHS in Scotland: The Legacy of the Past and the Prospect of the Future* (Aldershot, 2000), pp. 79–105; pp. 91–95.

31. R. C. O. Matthews, C. H. Feinstein and J. C. Odling-Smee, *British Economic Growth, 1856–1973* (Oxford, 1982), p. 435, in G. Owen, *From Empire to Europe: The Decline and Revival of British Industry since the Second World War* (London, 1999), p. 31.

32. Glynn and Booth, *Modern Britain*, p. 242.

33. The ongoing nature of many industrial illnesses is stressed in Tweedale, *Magic Mineral to Killer Dust*; Johnston and McIvor, *Lethal Work*; McIvor and Johnston, *Miner's Lung*.

34. M. Bunting, *Willing Slaves: How the Overwork Culture is Ruling Our Lives* (London, 2004).
35. Ibid, p. 77.
36. Ibid, p. 179.
37. Widely reported, for example www.guardian.co.uk/business/2008/jun/16/ primark.child.labour, consulted on 22 July 2008.
38. Labour Behind the Label campaign against the 'sweatshop' conditions in which it claims many clothes sold by British companies are made: www. labourbehindthelabel.org/.
39. C. Sellers, 'Cross-Nationalizing the History of Industrial Hazard', plenary lecture delivered at 'Environment, Health and History', London School of Hygiene and Tropical Medicine, 15 September 2007.
40. Johnston and McIvor, 'Whatever Happened to the *Occupational* Health Service?'.
41. C. Black, *Working for a Healthier Tomorrow: Review of the Health of Britain's Working Age Population* (London, 2008).
42. *High Quality Care for All: NHS Next Stage Review Final Report* (2008), Cm 7432.
43. Ibid, p. 9.
44. Black, *Working for a Healthier Tomorrow*, p. 10.
45. Ibid, p. 10. As this study has demonstrated, however, historically it has proven difficult to measure and quantify the benefits of health; employers have also historically shown a marked reluctance to invest in health improvement schemes for their employees, even when the savings in terms of lost working hours have been pointed out.
46. Ibid, p. 15.
47. Similar ideas underpin the CAPRIGHT project, a Europe wide initiative which seeks to apply Amartya Sen's concept of capability to align social justice and economic goals in employment policy: www.capright.eu.
48. Black, *Working for a Healthier Tomorrow*, p. 13. Black recognised that this 'health problem' was the one most in need of attention and acknowledged that the government's Pathways to Work scheme had largely failed those with mental health problems. Given the surge in work-related stress, however, it is far from clear that the Review addresses the characteristics of workplace practice which undermine mental well-being.
49. Ibid, p. 12. This approach mirrors strategies adopted by psychiatric social workers in the 1940s and 1950s who pioneered new ways to enable people with mental health problems to work by focussing on what they could do rather than their remaining symptoms: see V. Long, ' "A Satisfactory Job is the Best Psychotherapist: Employment and Mental Health, 1939–60', in J. Melling and P. Dale (eds), *Mental Illness and Learning Disability Since 1850: Finding a Place for Mental Disorder in the United Kingdom* (London, 2006), pp. 179–199.
50. Black, *Working for a Healthier Tomorrow*, p. 12.
51. Ibid, p. 12.

Bibliography

Primary Sources

Archives

Modern Records Centre, University of Warwick:
    Trades Union Congress Archive
    British Employers' Confederation Archive
    National Union of Railwaymen Archive
The National Archives
    Cabinet Office
    Prime Minister's Office
    Ministry of Labour
    Ministry of Health
    Public Records Office
Archive and Records Centre, D122, Boots UK, Nottingham
    Boots Company Archive
Birmingham Archives and Heritage, Central Library
    Cadbury Family Archives

Government Publications

Annual Reports of the Chief Inspector of Factories.
Departmental Committee on Lighting in Factories, *Lighting in Factories: Fifth Report* (London, 1940).
High Quality Care for All: NHS Next Stage Review Final Report (2008), Cm 7432.
Home Office and Scottish Home Department, *Health, Welfare, and Safety in Non-Industrial Employment: Hours of Employment of Juveniles* (1949), Cmd 7664.
Reports of the Industrial Fatigue Research Board.
Reports of the Industrial Health Research Board.
Industrial Health: A Survey of Halifax. Report by Her Majesty's Factory Inspectorate and Recommendations of the Industrial Health Advisory Committee (London, 1958).
Industrial Health: A Survey of the Pottery Industry in Stoke-on-Trent (London, 1959).
Ministry of Health and Department of Health for Scotland, *A National Health Service* (1944), Cmd 6502.
Ministry of Labour and National Service, 'Welfare Memorandum 3: Rest Breaks for Women and Girl Workers arranged by the Liverpool Union of Girls' Clubs', 14 November 1941.
Ministry of Labour and National Service Circular 128/87, 'Workers' Welfare: Holidays in Industry in 1942', 21 April 1942.
Ministry of Labour and National Service, *Memorandum on Medical Supervision in Factories Form 327* (London, 1946).

Ministry of Labour and National Service, *Fighting Fit in the Factory* (London, no date).
Ministry of Munitions, *Health of Munition Workers Committee Interim Report: Industrial Efficiency and Fatigue* (1917), Cd 8511.
Ministry of Munitions, *Health of Munition Workers Committee Final Report: Industrial Health and Efficiency* (1918), Cd 9065.
Report of a Committee of Enquiry on Industrial Health Services (1951), Cmd 8170.
Safety and Health at Work (1972), Cmnd 5034.
Select Committee on National Expenditure, *Third Report: Health and Welfare of Women in War Factories* (London, 1942).

Journals and Newspapers
British Journal of Industrial Medicine
British Medical Journal
Cavalcade
Hansard
Industrial Welfare (formerly known as the *Boys' and Industrial Welfare Journal* and the *Journal of Industrial Welfare*)
Justice
Municipal Engineering, Sanitary Record and Municipal Motor
Public Health
The Lancet
The Times
Journal of the Society of Occupational Medicine (formerly known as *Transactions of the Association of Industrial Medical Officers* and *Transactions of the Society of Occupational Medicine*)

Books and Reports
Adorno, T. W. and Horkheimer, M., *Dialectic of Enlightenment* (1944, trans. J. Cumming 1972, London, 1989).
Baker, C. W., *Government Control and Operation of Industry in Great Britain and the United States during the World War* (Oxford, 1921).
Beales, H. L. and Lambert, R. S. (eds), *Memoirs of the Unemployed* (London, 1934).
Bevin, E., 'First Meeting of the Industrial Health Advisory Committee: Speech Delivered by the Rt Hon Ernest Bevin MP' (London, 1943).
British Medical Association, *Report of Committee on Nutrition* (London, 1933).
British Medical Association, *Committee on Industrial Health in Factories* (1941).
British Medical Association, *The Future of Occupational Health Services* (London, 1961).
Collier, D. J., *The Girl in Industry* (London, 1918).
Drever, J., 'The Human Factor in Industrial Relations', in C. S. Myers (ed.), *Industrial Psychology* (2nd edition, Oxford, 1944: first published 1929), pp. 16–38.
Gollan, J., *Youth in British Industry: A Survey of Labour Conditions To-day* (London, 1937).
Hope, E. W. in collaboration with Hanna, W. and Stallybrass, C. O., *Industrial Hygiene and Medicine* (London, 1923).

Howard, E., *To-morrow: a Peaceful Path to Real Reform* (London, 1898).

Jahoda, M., Lazarsfeld, P. F. and Zeisel, H., *Marienthal: The Sociology of an Unemployed Community* (London, 1974. First published as *Die Arbeitslosen von Marienthal* in 1933).

Jenson and Nicholson Ltd, *The Function of Colour in Factories Schools and Hospitals* (London, n.d., circa 1940s?).

Kelly, E. T. (ed.), *Welfare Work in Industry – By Members of the Institute of Welfare Workers* (London, 1925).

Lee, J., *The Principles of Industrial Welfare* (London, 1924).

Legge, T., *Industrial Maladies* (Oxford, 1934).

MacDonald, S., *Simple Health Talks with Women War Workers* (London, 1917).

Martindale, H., *From One Generation to Another 1839–1944: A Book of Memoirs* (London, 1944).

Medical Services Review Committee, *A Review of the Medical Services in Great Britain* (London, 1962).

Milne-Bailey, W., *Trade Unions and the State* (London, 1934).

Mass Observation, *War Factory: A Report* (London, 1943).

Newman, G., *The Rise of Preventive Medicine* (Oxford, 1932).

Oastler, R., *The Factory Question: The Law or the Needle* (London, 1836).

Oliver, T., *The Health of Workers* (London, 1925).

Proud, D., *Welfare Work: Employers' Experiments for Improving Working Conditions in Factories with Foreword by David Lloyd George PM and first Minister of Munitions* (3rd edition, London, 1918).

Priestley, J. B., *English Journey: Being a Rambling but Truthful Account of What One Man Saw and Heard and Felt and Thought During a Journey Through England During the Autumn of the Year 1933, Introduced by Margaret Drabble* (1934; London, 1997).

Sigerist, H., *Socialised Medicine in the Soviet Union* (London, 1937).

Squire, R., *Thirty Years in the Public Service: An Industrial Retrospect* (London, 1927).

Trades Union Congress, *Reports of the Proceedings of the Forty-Sixth Annual Trades Union Congress* (London, 1913).

Turner Thackrah, C., *The Effects of the Principle Arts, Trades and Professions, and of Civic States and Habits of Living, on Health and Longevity* (London, 1831).

Ure, A., *The Philosophy of Manufactures: or, An Exposition of the Scientific, Moral, and Commercial Economy of the Factory System of Great Britain* (2nd edition, London, 1835).

Webb, B., *Health of Working Girls: A Handbook for Welfare Supervisors and Others* (London, 1917).

Secondary Sources

Adams, A., *Medicine by Design: the Architect and the Modern Hospital, 1893–1943* (Minneapolis, 2008).

Barlow, K., 'The Peckham Experiment', *Medical History*, 29 (1985), 264–271.

Barton, S., *Working-Class Holidays and Popular Tourism, 1840–1970* (Manchester, 2005).

Bartrip, P. W. J. and Burman, S. B., *The Wounded Soldiers of Industry: Industrial Compensation Policy 1833–1897* (Oxford, 1983).

Bartrip, P. W. J., *Workmen's Compensation in Twentieth Century Britain: Law, History and Social Policy* (Aldershot, 1987).

Bartrip, P., *Themselves Writ Large: The BMA 1832–1966* (London, 1996).

Bartrip, P., *The Home Office and the Dangerous Trades: Regulating Occupational Disease in Victorian and Edwardian Britain* (Amsterdam and New York, 2002).

Berg, M., *The Age of Manufactures 1700–1820: Industry, Innovation and Work in Britain* (London, 1994).

Black, C., *Working for a Healthier Tomorrow: Review of the Health of Britain's Working Age Population* (London, 2008).

Borowy, I. and Gruner, W. D. (eds), *Facing Illness in Troubled Times: Health in Europe in the Interwar Years 1918–1939* (Frankfurt am Main, 2005).

Bourdieu, P., *The Logic of Practice* (Cambridge, 1990, trans. R. Nice).

Bowden, S. and Tweedale, G., 'Mondays without Dread: The Trade Unions' Response to Byssinosis in the Lancashire Cotton Industry in the Twentieth Century', *Social History of Medicine*, 16 (2003), 79–96.

Braybon, G., *Women Workers in the First World War* (London, 1981).

Briggs, L., *The Rational Factory: Architecture, Technology and Work in America's Age of Mass Production* (Baltimore and London, 1996).

Bufton, M. W. and Melling, J., ' "A Mere Matter of Rock": Organised Labour, Scientific Evidence and British Government Schemes for Compensation of Silicosis and Pneumoconiosis among Coalminers, 1926–1940', *Medical History*, 49 (2005), 155–178.

Bufton, M. W. and Melling, J., 'Coming Up for Air: Experts, Employers, and Workers in Campaigns to Compensate Silicosis Sufferers in Britain, 1918–39', *Social History of Medicine*, 18 (2005), 63–86.

Bunting, M., *Willing Slaves: How the Overwork Culture is Ruling Our Lives* (London, 2004).

Buxton, N. K. and Aldcroft, D. H. (eds), *British Industry Between the Wars: Instability and Industrial Development 1919–1939* (London, 1979).

Calhoun, C. (ed.), *Habermas and the Public Sphere* (Massachusetts, 1999).

Carter, J. T., 'Fifty Years of Medicine in the Workplace: The Contribution of the Society of Occupational Medicine to Occupational Health Practice, 1935–1985', *Journal of the Society of Occupational Medicine*, 35 (1985), 4–22.

Carter, J. T., 'Twenty-One Years of EMAS', *Occupational Medicine*, 44 (1994), 119–122.

Carter, S., *Rise and Shine: Sunlight, Technology and Health* (Oxford, 2007).

Chance, H., 'The Angel in the Garden Suburb: Arcadian Allegory in the "Girls' Grounds" at the Cadbury Factory, Bournville, England, 1880–1930', *Studies in the History of Gardens and Designed Landscapes*, 27 (2007), 197–216.

Chin, C., *The Cadbury Story: A Short History* (Studley, 1998).

Cooter, R., 'The Meaning of Factures: Orthopaedics and the Reform of British Hospitals in the Inter-War Period', *Medical History*, 31 (1987), 306–332.

Cooter, R. and Pickstone, J. (eds), *Companion to Medicine in the Twentieth Century* (London, 2003).

Cooter, R., 'The Rise and Decline of the Medical Member: Doctors and Parliament in Edwardian and Interwar Britain', *Bulletin for the History of Medicine*, 78 (2004), 59–107.

Crawford, M., *Building the Workingman's Paradise: The Design of American Company Towns* (London, 1995).

Crooks, E., *The Factory Inspectors: A Legacy of the Industrial Revolution* (Stroud, 2005).

Crossley, N. and Roberts, J. M. (eds), *After Habermas: New Perspectives on the Public Sphere* (Oxford, 2004).

Crossley, N., *Contesting Psychiatry: Social Movements in Mental Health* (London, 2006).

Cumming, E. and Kaplan, W., *The Arts and Crafts Movement* (London, 1991).

Darley, G., *Factory* (London, 2003).

Davin, A., 'Imperialism and Motherhood', *History Workshop Journal*, 5 (1978), 9–66.

Dingwell, R., Rafferty, A. M. and Webster, C., *An Introduction to the Social History of Nursing* (London, 1988).

Dwork, D., *War is Good for Babies and Other Young Children: A History of the Infant and Child Welfare Movement in England 1898–1918* (London, 1989).

Earwicker, R., 'A Study of the BMA-TUC Joint Committee on Medical Questions, 1935–1939', *Journal of Social Policy*, 8 (1979), 335–561.

Edgerton, D., *Warfare State: Britain, 1920–1970* (Cambridge, 2006).

Engel, H., 'Why I Became an Occupational Physician', *Occupational Medicine*, 58 (2009), 151.

Finlayson, G., *Citizen, State, and Social Welfare in Britain, 1830–1990* (Oxford, 1994).

Fitzgerald, D., *Every Farm a Factory: The Industrial Ideal in American Agriculture* (New Haven, 2003).

Fitzgerald, R., *British Labour Management and Industrial Welfare, 1946–1939* (London, 1988).

Fraser, D., *The Evolution of the British Welfare State* (2nd edition, London, 1992).

Geraghty, T. G., 'Factory System', in J. Mokyr (ed.), *Oxford Encyclopaedia of Economic History Vol. 2* (Oxford, 2003), 247–253.

Gijswijt-Hofstra, M. and Marland, H. (eds), *Cultures of Child Health in Britain and the Netherlands in the Twentieth Century* (Amsterdam and Atlanta, 2003).

Glucksman, M., *Women Assemble: Women Workers and the New Industries in Inter-War Britain* (London, 1990).

Glynn, S. and Booth, A., *Modern Britain: An Economic and Social History* (London, 1996).

Gordon, C., *Dead on Arrival: The Politics of Health Care in Twentieth-Century America* (Princeton, 2005).

Gray, R., *The Factory Question and Industrial England, 1830–1860* (Cambridge, 1996).

Greenberg, M., 'The Last Senior Medical Inspector of Factories and His Place in the History of Occupational Health', *American Journal of Industrial Medicine*, 49 (2006), 54–9.

Greenlees, J., ' "Stop Kissing and Steaming!" Tuberculosis and the Occupational Health Movement in Massachusetts and Lancashire, 1870–1918', *Urban History*, 32 (2005), 223–246.

Guillén, M. F., *The Taylorized Beauty of the Mechanical: Scientific Management and the Rise of Modernist Architecture* (Princeton, 2006).

Gurney, P., 'Labour's Great Arch: Cooperation and Cultural Revolution in Britain, 1795–1926', in E. Furlough and C. Strikwerda (eds), *Consumers Against Capitalism? Consumer Cooperation in Europe, North America, and Japan 1848–1990* (Oxford, 1999), pp. 135–171.

Habermas, J., *The Structural Transformation of the Public Sphere: An Inquiry into a Category of Bourgeois Society* (trans. Thomas Burger, London, 1989).

Hamish Fraser, W., *A History of British Trade Unionism 1700–1998* (London, 1999).

Harrison, B., 'Suffer the Working Day – Women in the "Dangerous Trades," 1880–1914', *Women's Studies International Forum*, 13 (1990) 79–90.

Harrison, B., *Not only the 'Dangerous Trades': Women's Work and Health in Britain, 1880–1914* (Abingdon, 1996).

Harrison, J. F. C., *Robert Owen and the Owenites in Britain and America: The Quest for the New Moral World* (2nd edition, Aldershot, 1994).

Harvey, C. and Press, J., 'John Ruskin and the Ethical Foundations of Morris & Company, 1861–96', *Journal of Business Ethics*, 14 (1995), 181–194.

Hayes, N., 'More Than "Music While You Eat"? Factory and Hostel Concerts, "Good Culture" and the Workers', in N. Hayes and J. Hill (eds), *Millions Like Us? British Culture in the Second World War* (Liverpool, 1999), pp. 209–235.

Hayes, N., 'Did Manual Workers Want Industrial Welfare? Canteens, Latrines and Masculinity on British Building Sites, 1918–1970', *Journal of Social History*, 35 (2002), 637–658.

Hendrick, H., *Images of Youth: Age, Class and the Male Youth Problem, 1880–1920* (Oxford, 1990).

Hilton, M., *Consumerism in Twentieth-Century Britain* (Cambridge, 2003).

Hinton, J., *The First Shop Stewards' Movement* (London, 1973).

Hinton, J., *Labour and Socialism: A History of the British Labour Movement 1867–1974* (Brighton, 1983).

Johnston, R. and McIvor, A., *Lethal Work: A History of the Asbestos Tragedy in Scotland* (East Linton, 2000).

Johnston, R. and McIvor, A., 'Whatever Happened to the *Occupational* Health Service? The NHS, the OHS and the Asbestos Tragedy on Clydeside', in C. Nottingham (ed.), *The NHS in Scotland: The Legacy of the Past and the Prospect of the Future* (Aldershot, 2000), pp. 79–105.

Johnston, R. and McIvor, A., 'The War and the Body at Work: Occupational Health and Safety in Scottish Industry, 1939–45', *Journal Of Scottish Historical Studies*, 24 (2004), 113–136.

Johnston, R. and McIvor, A., 'Marginalising the Body at Work? Employers' Occupational Health Strategies and Occupational Medicine in Scotland c. 1930–1974', *Social History of Medicine*, 21 (2008), 127–144.

Jones, E., *Industrial Architecture in Britain, 1750–1939* (London, 1985).

Jones, G., *Social Hygiene in Twentieth Century Britain* (London, 1986).

Jones, H., 'Employers' Welfare Schemes and Industrial Relations in Inter-War Britain', *Business History*, 25 (1983), 61–75.

Jones, H., 'Women Health Workers: The Case of the First Women Factory Inspectors in Britain', *Social History of Medicine*, 1 (1988), 165–181.

Joyce, P., *Work, Society and Politics: The Culture of the Factory in Later Victorian England* (London, 1982).

Lane, J., *A Social History of Medicine: Health, Healing and Disease in England, 1750–1950* (London, 2001).

Lewis, B., *'So Clean': Lord Leverhulme, Soap and Civilization* (Manchester, 2008).

Lewis, J., *The Politics of Motherhood: Child and Maternal Welfare in England 1900–1939* (London, 1980).

Lewis, J. and Brookes, B., 'The Peckham Health Centre, "PEP", and the Concept of General Practice during the 1930s and 1940s', *Medical History*, 27 (1983), 151–61.

Long, V., ' "A Satisfactory Job is the Best Psychotherapist": Employment and Mental Health, 1939–60', in P. Dale and J. Melling (eds), *Mental Illness and Learning Disability Since 1850: Finding a Place for Mental Disorder in the United Kingdom* (Abingdon, 2006).

Long, V. and Marland, H., 'From Danger and Motherhood to Health and Beauty: Health Advice for the Factory Girl in Early Twentieth-Century Britain', *Twentieth Century British History*, 20 (2009), 454–481.

Long, V., 'Industrial Homes, Domestic Factories: the Convergence of Public and Private Space in Interwar Britain', *Journal of British Studies*, 50 (2011).

Luckin, B., *Questions of Power: Electricity and Environment in Interwar Britain* (Manchester, 1990).

MacDonald, D. F, *The State and the Trade Unions* (2nd edition, London, 1976).

Malone, C., *Women's Bodies and Dangerous Trades in England 1880–1914* (Woodbridge, 2003).

Martin, R. M., *TUC: The Growth of a Pressure Group 1868–1976* (Oxford, 1980).

Matless, D., *Landscape and Englishness* (London, 1998).

Matless, D., 'Bodies Made of Grass Made of Earth made of Bodies: Organicism, Diet and National Health in Mid-Twentieth-Century England', *Journal of Historical Geography*, 27 (2001), 355–376.

Mayhew, M., 'The 1930s Nutrition Controversy', *Journal of Contemporary History*, 23 (1988), pp. 445–464.

McIvor, A. J., 'Employers, the Government, and Industrial Fatigue in Britain, 1890–1918', *British Journal of Industrial Medicine*, 44 (1987), 724–732.

McIvor, A., *Organised Capital: Employers' Associations and Industrial Relations in Northern England, 1880–1939* (Cambridge, 1996).

McIvor, A., *A History of Work in Britain, 1880–1950* (Hampshire, 2001).

McIvor, A. and Johnston, R., *Miners' Lung: A History of Dust Disease in British Coal Mining* (Aldershot, 2007).

McKibbin, R., *The Ideologies of Class: Social Relations in Britain 1880–1950* (London, 1991).

Meharg, A., 'The Arsenic Green', *Nature*, 423 (2003), p. 688.

Melling, J., 'Employers, Industrial Welfare, and the Struggle for Work-Place Control in British Industry, 1880–1920', in H. F. Gospel and C. R. Littler (eds), *Managerial Strategies and Industrial Relations: An Historical and Comparative Study* (London, 1983), pp. 55–81.

Melling, J., 'The Risks of Working and the Risks of Not Working: Trade Unions, Employers and Responses to the Risk of Occupational Illness in British Industry, c. 1890–1940s', *ESRC Centre for Analysis of Risk and Regulation Discussion Paper 12* (2003), pp. 14–34.

Middlemas, K., *Politics in Industrial Society: The Experience of the British System Since 1911* (London, 1979).

Murphy, S., 'The Early Days of the MRC Social Medicine Research Unit', *Social History of Medicine*, 12 (1999), 389–406.

Nye, D., *American Technological Sublime* (Cambridge, Massachusetts, 1994).

Oswald, N., 'Training Doctors for the National Health Service: Social Medicine, Medical Education and the GMC 1936–48', in D. Porter (ed.), *Social Medicine and Medical Sociology in the Twentieth Century* (Amsterdam, 1997), pp. 59–80.

Owen, G., *From Empire to Europe: The Decline and Revival of British Industry since the Second World War* (London, 1999).

Oxford English Dictionary (2nd edition, 1989).

Oxford Dictionary of National Biography (online edition, Oxford, 2004–2010).

Perkins, H., *The Rise of Professional Society: England Since 1880* (London, 1990).

Pick, D., *War Machine: The Rationalisation of Slaughter in the Modern Age* (Yale, 1993).

Porter, D., *Health, Civilisation and the State: A History of Public Health from Ancient to Modern Times* (London, 1999).

Price, R., and Bain, G. S., 'The Labour Force' in A. H Halsey (ed.), *British Social Trends Since 1900* (Basingstoke, 2000), pp. 162–201.

Richards, T., *The Commodity Culture of Victorian England: Advertising and Spectacle, 1851–1914* (Stanford, 1990).

Riley, J. C., *Sick, Not Dead: The Health of the British Workingmen During the Mortality Decline* (Baltimore, 1997).

Robertson, E., Pickering, M. and Korczynski, M., 'Harmonious Relations? Music at Work in the Rowntree and Cadbury Factories', *Business History*, 49 (2007), 211–234.

Robertson, E., *Chocolate, Women and Empire: A Social and Cultural History* (Manchester, 2009).

Rose, N., *Governing the Soul: The Shaping of the Private Self* (2nd edition, London, 1999).

Rubin, G. R., *War, Law and Labour: The Munitions Acts, State Regulation, and the Unions, 1915–21* (Oxford, 1987).

Scarry, E., *The Body in Pain: The Making and Unmaking of the World* (Oxford, 1985).

Scull, A., (ed.), *The Asylum As Utopia: W. A. F. Browne and the Mid-Nineteenth Century Consolidation of Psychiatry* (London, 1991).

Searle, G. R., *The Quest for National Efficiency* (Oxford, 1971).

Sellers, C., *Hazards of the Job: From Industrial Disease to Environmental Health Science* (Chapel Hill and London, 1997).

Skinner, J., *Form and Fancy: Factories and Factory Buildings by Wallis, Gilbert & Partners, 1916–1939* (Liverpool, 1997).

Smith, D. F., *Nutrition in Britain: Science, Scientists and Politics in the Twentieth Century* (London, 1997).

Stevenson, J., *The Penguin Social History of Britain: British Society 1914–45* (London, 1984).

Stewart, J., *The Battle for Health: A Political History of the Socialist Medical Association* (Aldershot, 1999).

Stratton, M. and Trinder, B., *Twentieth-Century Industrial Archaeology* (London, 2000).

Sturdy, S., 'Hippocrates and State Medicine: George Newman Outlines the Founding Policy of the Ministry of Health', in C. Lawrence and G. Weisz (eds), *Greater than the Parts: Holism in Biomedicine 1920–1950* (Oxford, 1998), pp. 112–134.

Sturdy, S. and Cooter, R., 'Science, Scientific Management, and the Transformation of Medicine in Britain c. 1870–1950', *History of Science*, 36 (1998), 1–47.

Sturdy, S. (ed.), *Medicine, Health and the Public Sphere in Britain 1600–2000* (London, 2002).

Summerfield, P., 'Women, Work and Welfare: A Study of Child Care and Shopping in Britain in the Second World War', *Journal of Social History*, 17 (1983), 249–269.

Taylor, R., *The TUC: From the General Strike to New Unionism* (Houndmills, 2000).

Thane, P., *Foundations of the Welfare State* (2nd edition, Harlow, 1996).

Thom, D., *Nice Girls and Rude Girls: Women Workers in World War 1* (2nd edition, London, 2000).

Thomson, M., *Psychological Subjects: Identity, Culture and Health in Twentieth-Century Britain* (Oxford, 2006).

Tiratsoo, N. and Tomlinson, J., *Industrial Efficiency and State Intervention: Labour 1939–51* (London, 1993).

Tomes, N., *The Gospel of Germs: Men, Women and the Microbe in American Life* (Cambridge, Massachusetts, 1998).

Tweedale, G., *Magic Mineral to Killer Dust: Turner and Newall and the Asbestos Hazard* (Oxford, 2000).

Vernon, J., *Hunger: A Modern History* (Cambridge and London, 2007).

Waldron, H. A., 'Occupational Health during the Second World War: Hope Deferred or Hope Abandoned?' *Medical History*, 41 (1997), 197–212.

Walton, J. K., 'Towns and Consumerism', in M. Daunton (ed.), *The Cambridge Urban History of Britain Volume III 1840–1950* (Cambridge, 2000), pp. 715–744.

Ward, S. V. (ed.), *The Garden City: Past, Present and Future* (London, 1992).

Webster, C., 'Healthy or Hungry Thirties?', *History Workshop Journal*, 13 (1982) 110–129.

Webster, C., *The Health Services Since the War Volume 1: Problems of Health Care – The National Health Service Before 1957* (London, 1988).

Weindling, P. (ed.), *The Social History of Occupational Health* (London, 1985).

Weiner, M. J., *English Culture and the Decline of the Industrial Spirit 1850–1980* (Cambridge, 1981).

Whiteside, N., 'Counting the Cost: Sickness and Disability among Working People in an Era of Industrial Recession, 1920–39', *Economic History Review*, 40 (1987), 228–246.

Williams, R., *The Country and the City* (London, 1973).

Winter, J. M., *The Great War and the British People* (2nd edition, Basingstoke, 2003).

Wohl, A. S., *Endangering Lives: Public Health in Victorian Britain* (London, 1984).

Wollacott, A., 'Maternalism, Professionalism and Industrial Welfare Supervisors in World War 1 Britain', *Women's History Review*, 3 (1994), 29–56.

Worster, D., *Nature's Economy: A History of Ecological Ideas* (2nd edition, Cambridge, 1994).

Wrightman, C., *More Than Munitions: Women, Work and the Engineering Industries 1900–1950* (Harlow, 1999).

Yeomans, D. and Cotton, D., *Owen Williams: The Engineer's Contribution to Contemporary Architecture* (London, 2001).

Zweiniger-Bargielowska, I., 'Raising a Nation of "Good Animals": The New Health Society and Health Education Campaigns in Interwar Britain', *Social History of Medicine*, 20 (2007), 73–89.

Unpublished Theses

Aspinall, S., 'Nature as well as Nurture: Environmentalism in Representations of Women and Exercise in Britain from the 1880s to the Early 1920s' (PhD thesis, University of Warwick, 2009).

Baillie, M., 'The Red Women of Clydeside: Women Munition Workers in the West of Scotland During the First World War' (PhD Thesis, McMaster University, 2002).

Bunce, J., 'The "Healthiest of Occupations" or One of the Most Dangerous? The Health Hazards of Agriculture in Britain 1934–2004' (MA thesis, University of Warwick, 2004).

Long, V., 'Changing Public Representations of Mental Illness in Britain, 1870–1970' (PhD thesis, University of Warwick, 2004).

Phillips, S., 'Industrial Welfare and Recreation at Boots Pure Drug Company 1883–1945' (PhD thesis, Nottingham Trent University, 2003).

Walker, D., 'Occupational Health and Safety in the British Chemical Industry, 1914–1974' (PhD Thesis, University of Strathclyde, 2007).

Internet Sources

Association of Group Occupational Health Services, http://www.agohs.org.uk/index.htm, accessed 31 July 2008.

Department for Work and Pensions, http://www.workingforhealth.gov.uk, consulted 29 April 2010.

Finch, J., 'Primark Axes Suppliers for using Child Labour', http://www.guardian.co.uk/business/2008/jun/16/primark.child.labour, 16 June 2008.

Labour Behind the Label, http://www.labourbehindthelabel.org/, consulted on 22 July 2008.

Wikipedia, 'William Morris', http://en.wikipedia.org/wiki/William_Morris, consulted on 1 February 2008.

Index